THE

ZOOLOGY

OF

THE VOYAGE OF H.M.S. BEAGLE,

UNDER THE COMMAND OF CAPTAIN FITZROY, R.N.,

DURING THE YEARS

1832 TO 1836.

PUBLISHED WITH THE APPROVAL OF
THE LORDS COMMISSIONERS OF HER MAJESTY'S TREASURY.

Edited and Superintended by
CHARLES DARWIN, ESQ. M.A. F.R.S. Sec. G.S.
NATURALIST TO THE EXPEDITION.

PART III.

BIRDS,

BY

JOHN GOULD, ESQ. F.L.S.

Copyright © 2018 Read Books Ltd.
This book is copyright and may not be
reproduced or copied in any way without
the express permission of the publisher in writing

British Library Cataloguing-in-Publication Data
A catalogue record for this book is available from
the British Library

Ornithology

Ornithology is a branch of zoology that concerns the study of birds. Etymologically, the word 'ornithology' derives from the ancient Greek ὄρνις *ornis* (bird) and λόγος *logos* (rationale or explanation). The science of ornithology has a long history and studies on birds have helped develop several key concepts in evolution, behaviour and ecology such as the definition of species, the process of speciation, instinct, learning, ecological niches and conservation. Whilst early ornithology was principally concerned with descriptions and distributions of species, ornithologists today seek answers to very specific questions, often using birds as models to test hypotheses or predictions based on theories. However, most modern biological theories apply across taxonomic groups, and consequently, the number of professional scientists who identify themselves as 'ornithologists' has declined. That this specific science has become part of the biological mainstream though, is in itself a testament to the field's importance.

Humans observed birds from the earliest times, and Stone Age drawings are among the oldest indications of an interest in birds, primarily due to their importance as a food source. One of the first key texts on ornithology was Aristotle's *Historia Animalium* (350 BC), in which he noted the habit of bird migration, moulting, egg laying and life span. He also propagated several, unfortunately false myths, such as the idea that swallows hibernated in winter. This idea became so well established, that even as late as 1878, Elliott Coues (an American surgeon, historian and ornithologist) could list as many as 182 contemporary publications dealing with the hibernation of swallows. In the Seventeenth century, Francis Willughby (1635–1672) and John Ray (1627–1705) came up with the first major system of bird classification that was based on function and morphology rather than on form or behaviour, this was a major breakthrough in terms of scientific thought, and Willughby's *Ornithologiae libri tres* (1676), completed by John Ray is often thought to mark the beginning of methodical ornithology. It was not until the Victorian era though, with the emergence of the gun and the concept of natural history, that ornithology emerged as a specialized science. This specialization led to the formation in Britain of the British Ornithologists' Union in 1858, and the following year, its journal *The Ibis* was founded.

This sudden spurt in ornithology was also due in part to colonialism. The bird collectors of the Victorian era observed the variations in bird forms and habits across geographic regions, noting local specialization and variation in widespread species. The collections of museums and private collectors grew with contributions from various parts of the world. This spread of the science meant that many amateurs became interested in 'bird watching' – with real possibilities to contribute knowledge. As early as 1916, Julian Huxley wrote a two part article in the *Auk*, noting the tensions between amateurs and professionals and suggesting that the 'vast army of bird-lovers and bird-watchers could begin providing the data scientists needed to address the fundamental problems of biology.' Organizations were started in many countries and these grew rapidly in membership, most notable among them being the Royal Society for the Protection of Birds (RSPB), founded in 1889 in Britain and the Audubon Society, founded in 1885 in the US.

Today, the science of ornithology is thriving, with many practical and economic applications such as the management of birds in food production (grainivorous birds, such as the Red billed Quelea are a major agricultural pest in parts of Africa), and the study of birds, as carriers of human diseases, such as Japanese Encephalitis, West Nile Virus, and H5N1. Of course, many species of birds have been driven to (or near) extinction by human activities, and hence ornithology has played an important part in conservation, utilising many location specific approaches. Critically endangered species such as the California Condor have been captured and bred in captivity, and it is hoped that many more birds can be saved in a like manner.

CORRIGENDA.

I am indebted to Mr. G. R. Gray for the following remarks and corrections:—

Page 13, to Milvago ocrocephalus, *Spix.* add
    Polyborus ocrocephalus, *Jard. & Selby's Ill.* t. 5.
Alter 7, 8, 9, & 10, to 5, 6, 7, & 8.
Page 15, Milvago leucurus, add
    Falco Australis, *Jard. & Selby's Ill. Orn.* n. s. pl. 24.
Page 49, Serpophaga, *Gould,* is probably synonymous with Euscarthmus, *Pr. Max.*
Page 56, Agriornis, *Gould,* is synonymous with Dasycephala of Swainson, and Tamnolanius, of Lesson; the species therefore should be
  sp. 1. D. lividus, *G. R. Gray.*
    Thamnophilus lividus, *Kittl. Voy. de Chili,* pl. 1.
    Tyrannus gutturalis, *Eyd. & Gerv. &c.*
  sp. 2. D. striata, *G. R. Gray.*
    Agr. striatus, *Gould.*
    Agr. micropterus, juv. *Gould,* sp. 3.
Page 57, sp. 4. D. maritima, *G. R. Gray.*
    Agr. maritimus, *G. R. Gray,* &c.
Page 66. The generic appellation of Opetiorhynchus, was adopted after the subjection of Mr. Gould; since its publication, however, I have considered that it might cause confusion with Furnarius, of Vieillot, as it is Temminck's name for the identical same division, therefore only a synonym, and am on that ground induced to change and propose the name of Cinclodes, which has been adopted by a Continental writer. The species should be altered thus:—

Page 66, Sp. 1. Cinclodes vulgaris, *G. R. Gray.*
Page 67, sp. 2. C. Patagonicus, *G. R. Gray, List of the Genera of Birds.*
  sp. 3. C. antarcticus, *G. R. Gray.*
    Cinclodes fuliginosus, *Less.*
Page 68, sp. 4. C. nigrofumosus, *G. R. Gray.*
Page 69, Eremobius, being previously employed, it is changed to Enicornis, *G. R. Gray.* The species to
  En. phœnicurus, *G. R. Gray, List of the Genera of Birds.*
Page 70, Rhinomya, being also previously employed; it is therefore changed to Rhinocrypta, *G. R. Gray.* The species to
  R. lanceolata, *G. R. Gray.*
Page 76, for Synallaxis major, *Gould,* read Anumbius acuticaudatus, *G. R. Gray.*
  Furnarius annumbi, *Vieill.*
  L'Anumbi, *Azara,* No. 222.
  Anthus acuticaudatus, *Less.*
  Anumbius anthoides, *D'Orb. & Lefr.*
Page 94, Fringilla fruticeti, *Kittl.* gives place to
  Fringilla erythrorhyncha, *Less. Voy. Thetis.* ii. p. 324.

LIST OF PLATES.

Plate I.	Milvago albogularis.		Plate XXV.	Limnornis curvirostris.
II.	Craxirex Galapagoensis.		XXVI.	———— rectirostris.
III.	Otus Galapagoensis.		XXVII.	Dendrodamus leucosternus.
IV.	Strix punctatissima.		XXVIII.	Sylvicola aureola.
V.	Progne modestus.		XXIX.	Ammodramus longicaudatus.
VI.	Pyrocephalus parvirostris.		XXX. {	Ammodramus Manimbè, *in place of* Ammodramus xanthornus.
VII.	———— nanus.			
VIII.	Tyrannula magnirostris.		XXXI.	Passer Jagoensis.
IX.	Lichenops erythropterus.		XXXII.	Chlorospiza melanodera.
X.	Fluvicola Azaræ.		XXXIII.	———— xanthogramma.
XI. {	Xolmis variegata, *in place of* Tænioptera variegata.		XXXIV. {	Aglaia striata, *in place of* Tanagra Darwinii.
XII.	Agriornis micropterus.		XXXV.	Pipilo personata.
XIII.	———— leucurus.		XXXVI.	Geospiza magnirostris.
XIV.	Pachyramphus albescens.		XXXVII.	———— strenua.
XV.	———— minimus.		XXXVIII.	———— fortis.
XVI.	Mimus trifasciatus.		XXXIX.	———— parvula.
XVII.	———— melanotis.		XL.	Camarhynchus psittaculus.
XVIII.	———— parvulus.		XLI.	———— crassirostris.
XIX. {	Uppucerthia dumetoria, *in place of* Upercerthia dumetaria.		XLII.	Cactornis scandens.
			XLIII.	———— assimilis.
XX. {	Opetiorhynchus nigrofumosus, *in place of* Opetiorhynchus lanceolatus.		XLIV.	Certhidea olivacea.
			XLV.	Xanthornus flaviceps.
			XLVI.	Zenaida Galapagoensis.
XXI.	Eremobius phœnicurus.		XLVII.	Rhea Darwinii.
XXII. {	Anumbius acuticaudatus, *in place of* Synallaxis major.		XLVIII.	Zapornia notata.
			XLIX.	———— spilonota.
XXIII.	Synallaxis rufogularis.		L.	Anser melanopterus.
XXIV.	———— flavogularis.			

ADVERTISEMENT.

When I presented my collection of Birds to the Zoological Society, Mr. Gould kindly undertook to furnish me with descriptions of the new species and names of those already known. This he has performed, but owing to the hurry, consequent on his departure for Australia,—an expedition from which the science of Ornithology will derive such great advantages,—he was compelled to leave some part of his manuscript so far incomplete, that without the possibility of personal communication with him, I was left in doubt on some essential points. Mr. George Robert Gray, the ornithological assistant in the Zoological department of the British Museum, has in the most obliging manner undertaken to obviate this difficulty, by furnishing me with information with respect to some parts of the general arrangement, and likewise on that most intricate subject,—the knowledge of what species have already been described, and the use of proper generic terms. I shall endeavour in every part of the text to refer to Mr. G. R. Gray's assistance, where I have used it. As some of Mr. Gould's descriptions appeared to me brief, I have enlarged them, but have always endeavoured to retain his specific character; so that, by this means, I trust I shall not throw any obscurity on what he considers the essential character in each case; but at the same time, I hope, that these additional remarks may render the work more complete.

The accompanying illustrations, which are fifty in number, were taken from sketches made by Mr. Gould himself, and executed on stone by Mrs. Gould, with that admirable success, which has attended all her works. They are all of the natural size with the exception of four raptorial birds, a goose and a species of Rhea. As the dimensions of these latter birds are given, their proportional reduction will readily be seen. I had originally intended to have added the initial letter of my name to the account of the habits and ranges, and that of Mr. Gould's to the description of the genera and species; but as it may be known that he is responsible for the latter, and myself for the former, this appeared to me useless; and I have, therefore, thought it better to incorporate all general remarks in my own name, stating on every occasion my authority, so that wherever the personal pronoun is used it refers to myself. Finally, I must remark, that after the excellent dissertation, now in the course of publication, on the habits and distribution of the birds of South America by M. Alcide D'Orbigny, in which he has combined his own extended observations with those of Azara, my endeavour to add anything to our information on this subject, may at first be thought superfluous. But as during the Beagle's voyage, I visited some portions of America south of the range of M. D'Orbigny's travels, I shall relate in order the few facts, which I have been enabled to collect together; and these, if not new, may at least tend to confirm former accounts. I have, however, thought myself obliged to omit some parts, which otherwise I should have given; and, after having read the published portion of M. D'Orbigny's great work, I have corrected some errors, into which I had fallen. I have not, however, altered any thing simply because it differs from what that gentleman may have written; but only where I have been convinced that my means of observation were inferior to his.

BIRDS.

Family—VULTURIDÆ.

Sarcoramphus gryphus. *Bonap.*

Vultur gryphus, *Linn.*
—————, *Humb.* Zoolog. p. 31.
Sarcoramphus Condor, *D'Orbigny.* Voy. Ois.
Condor of the inhabitants of South America.

The Condor is known to have a wide range, being found on the west coast of South America, from the Strait of Magellan, throughout the range of the Cordillera, as far, according to M. D'Orbigny, as 8° north latitude. On the Patagonian shore, the steep cliff near the mouth of the Rio Negro, in latitude 41°, was the most northern point where I ever saw these birds, or heard of their existence; and they have there wandered about four hundred miles from the great central line of their habitation in the Andes. Further south, among the bold precipices which form the head of Port Desire, they are not uncommon; yet only a few stragglers occasionally visit the sea-coast. A line of cliff near the mouth of the Santa Cruz is frequented by these birds, and about eighty miles up the river, where the sides of the valley were formed by steep basaltic precipices, the Condor again appeared, although in the intermediate space not one had been seen. From these and similar facts, I believe that the presence of this bird is chiefly determined by the occurrence of perpendicular cliffs. In Patagonia the Condors, either by pairs or many together, both sleep and breed on the same overhanging ledges. In Chile, however, during the greater part of the year, they haunt the lower country, near the shores of the Pacific, and at night several roost in one tree; but in the early part of summer they retire to the most inaccessible parts of the inner Cordillera, there to breed in peace.

With respect to their propagation, I was told by the country people in Chile, that the Condor makes no sort of nest, but in the months of November and December, lays two large white eggs on a shelf of bare rock. Certainly, on the Patagonian coast, I could not see any sort of nest among the cliffs, where the young ones were standing. I was told that the young Condors could not fly for a whole year, but this probably was a mistake, since M. D'Orbigny says they take to the wing in about a month and a half after being hatched. On the fifth of March (corresponding to our September), I saw a young bird at Concepcion, which, though in size only little inferior to a full-grown one, was completely covered by down, like that of a gosling, but of a blackish colour. I can, however, scarcely believe that this bird could have used, for some months subsequently, its wings for flight. After the period when the young Condor can fly, apparently as well as the old birds, they yet remain (as I observed in Patagonia) both roosting at night on the same ledge, and hunting by day with their parents: but before the young bird has the ruff round its neck white, it may often be seen hunting by itself. At the mouth of the Santa Cruz, during part of April and May, a pair of old birds might be seen every day, either perched on a certain ledge, or sailing about in company with a single young one, which latter, though full fledged, had not its ruff white.

The Condors generally live by pairs; but among the basaltic cliffs of the plains, high up the river Santa Cruz, I found a spot where scores must usually haunt. They were not shy; and on coming suddenly to the brow of the precipice, it was a fine sight to see between twenty and thirty of these great* birds start heavily from their resting place, and wheel away in majestic circles. From the large quantity of dung on the rocks, they must have long frequented this cliff; and probably they both roost and breed there. Having gorged themselves with carrion on the plains below, they retire to these favourite ledges to digest their food in quietness. From these facts, the Condor must, to a certain degree be considered, like the Gallinazo (*Cathartes atratus*), a gregarious bird. In this part of the country they live almost entirely on the guanacoes, which either have died a natural death, or, as more commonly happens, have been killed by the pumas. I believe, from what I saw in Patagonia, that they do not, on ordinary occasions, extend their daily excursions to any great distance from their regular sleeping places.

The condors may oftentimes be seen at a great height, soaring over a certain spot in the most graceful spires and circles. On some occasions I am sure that they do this for their sport; but on others, the Chileno countryman tells you, that they are watching a dying animal, or the puma devouring its prey. If the condors

---

* I measured a specimen, which I killed there: it was from tip to tip of wing, eight and a half feet; and from end of beak to end of tail four feet.

glide down, and then suddenly all rise together, the Chileno knows that it is the puma, which, watching the carcass, has sprung out to drive away the robbers. Besides feeding on carrion, the condors frequently attack young goats and lambs. Hence the shepherds train their dogs, the moment the enemy passes over, to run out, and looking upwards, to bark violently. The Chilenos destroy and catch numbers; two methods are used: one is to place a carcass within an enclosure of sticks on a level piece of ground, and when the condors have gorged themselves to gallop up on horseback to the entrance, and thus enclose them: for when this bird has not space to run, it cannot give its body sufficient momentum to rise from the ground. The second method is to mark the trees in which, frequently to the number of five or six, they roost together, and then at night to climb up and noose them; they are such heavy sleepers, as I have myself witnessed, that this is not a difficult task. At Valparaiso I have seen a living condor sold for sixpence, but the common price is eight or ten shillings. One which I saw brought in for sale, had been lashed with a rope, and was much injured; but the moment the line was cut by which its bill was secured, it began, although surrounded by people, ravenously to tear a piece of carrion. In a garden at the same place, between twenty and thirty of these birds were kept alive; they were fed only once a week, yet they appeared to be in pretty good health.* The Chileno countrymen assert, that the condor will live and retain its powers between five and six weeks without eating: I cannot answer for the truth of this fact, but it is a cruel experiment, which very likely has been tried.

When an animal is killed in this country, it is well known that the condors, like other carrion vultures, gain the intelligence and congregate in a manner which often appears inexplicable. In most cases, it must not be overlooked, that the birds have discovered their prey, and have picked the skeleton clean, before the flesh is in the least degree tainted. Remembering the opinion of M. Audubon on the deficient smelling powers of such birds,† I tried in the above mentioned garden, the following experiment. The condors were tied, each by a rope, in a long row at the bottom of a wall. Having folded a piece of meat in white paper, I walked backwards and forwards, carrying it in my hand at the

* I noticed that several hours before any of the Condors died, all the lice with which they are infested, crawled to the outside feathers. I was told, that this always happened.

† In the case of the *Cathartes Aura*, Mr. Owen, in some notes read before the Zoological Society, (See Magazine of Nat. Hist. New Ser. vol. i. p. 638.) has demonstrated from the developed form of the olfactory nerves, that this bird must possess an acute sense of smell. It was mentioned on the same evening, in a communication from Mr. Sells, that on two occasions, persons in the West Indies having died, and their bodies not being buried till they smelt offensively, these birds congregated in numbers on the roof of the house. This instance appears quite conclusive, as it was certain, from the construction of the buildings, that they must have gained the intelligence by the sense of smell alone, and not by that of sight. It would appear from the various facts recorded, that carrion-feeding hawks possess both senses, in a very high degree.

distance of about three yards from them; but no notice whatever was taken of it. I then threw it on the ground within one yard of an old cock bird; he looked at it for a moment with attention, but then regarded it no more. With a stick I pushed it closer and closer, until at last he touched it with his beak: the paper was then instantly torn off with fury, and at the same moment every bird in the long row began struggling and flapping its wings. Under the same circumstances, it would have been quite impossible to have deceived a dog.

When the condors in a flock are wheeling round and round any spot, their flight is beautiful. Except when they rise from the ground, I do not recollect ever to have seen one flap its wings. Near Lima, I watched several of these birds for a quarter and half-an-hour, without once taking off my eyes. They moved in large curves, sweeping in circles, descending and ascending without once flapping. As several glided close over my head, I intently watched, from an oblique position, the separate and terminal feathers of the wing; if there had been the least vibratory movement, their outlines would have been blended together, but they were seen distinct against the blue sky. The head and neck were moved frequently, and apparently with force. If the bird wished to descend, the wings were for a moment collapsed; and then, when again expanded with an altered inclination, the momentum gained by the rapid descent, seemed to urge the bird upwards, with the even and steady movement of a paper kite. It was a beautiful spectacle thus to behold these great vultures hour after hour, without any apparent exertion, wheeling and gliding over mountain and river.

In the garden at Valparaiso, where so many condors were kept alive, I observed that all the hens had the iris of their eyes bright red, but the cocks yellowish-brown. In a young bird, whose back was brown, and ruff not white, (but which must have been at least nearly a year old, as it was then the spring) I observed that the eye was dark brown: upon examination after death, this proved to be a female, and therefore I suppose the colour of the iris changes at the same time with the plumage.

1. CATHARTES ATRATUS. *Rich. and Swain.*

Cathartes urubu, *D'Orbigny.* Voy. Ois.
Vultur atratus, *Bartram,* p. 287.
―――― jota, *Jardine's* Wilson, vol. iii. p. 236.
――――――, *Bonaparte's* List, p. 1.
Gallinazo or Cuervo of the Spanish inhabitants of America; and Black Vulture or Carrion Crow of the English of that continent.

THESE birds, I believe, are never found further south, than the neighbourhood of the Rio Negro, in latitude 41°: I never saw one in southern Patagonia, or in Tierra del Fuego. They appear to prefer damp places, especially the vicinity of rivers; and thus, although abundant both at the Rio Negro and Colorado, they are not found on the intermediate plains. Azara* states, that there existed a tradition in his time, that on the first arrival of the Spaniards in the Plata, these birds were not found in the neighbourhood of Monte Video, but that they subsequently followed the inhabitants from more northern districts. M. Al. D'Orbigny, in reference to this statement, observes that these vultures, although common on the northern bank of the Plata, and likewise on the rivers south of it, are not found in the neighbourhood of Buenos Ayres, where the immense slaughtering establishments are attended by infinite numbers of Polybori and gulls. M. D'Orbigny supposes that their absence is owing to the scarcity of trees and bushes in the Pampas; but this view, I think, will hardly hold good, inasmuch as the country near Bahia Blanca, where the Gallinazo (together with the carrion-feeding gull) is common, is as bare, if not more so, than the plains near Buenos Ayres. I have never seen the Gallinazo in Chile; and Molina, who was aware of the difference between the *C. atratus* and *C. aura,* has not noticed it; yet, on the opposite side of the Cordillera, near Mendoza, it is common. They do not occur in Chiloe, or on the west coast of the continent south of that island. In Wilson's Ornithology it is said that "the carrion crow (as this bird is called in the United States) is seldom found on the Atlantic to the northward of Newbern, lat. 35° North Carolina." But in Richardson's "Fauna Boreali-Americana," it is mentioned, on the authority of Mr. David Douglas, that on the Pacific side of the continent, it is common on the marshy islands of the Columbia, and in the neighbourhood of Lewis's and Clark's rivers (45°―47° N.) It has, therefore, a wider range in the northern

* Voyage dans l'Amérique Méridionale, vol. iii. p. 24.

than in the southern half of the continent. These vultures certainly are gregarious; for they seem to have pleasure in each other's society, and are not solely brought together by the attraction of a common prey. On a fine day, a flock may often be seen at a great height; each bird wheeling round and round in the most graceful evolutions. This is evidently done for their sport; or, perhaps, is connected (for a similar habit may sometimes be observed during the breeding season amongst our common rooks) with their matrimonial alliances.

2. CATHARTES AURA. *Illi.*

Vultur aura, *Linn.*
——————, *Jardine's Wilson*, vol. iii. p. 226.
Vultur jota, *Molina*, Compendio de la Hist. del Reyno de Chile, vol i. p. 296.
Turkey-buzzard and Carrion Crow of the English in America.

THIS bird has a wide geographical range, being found from 55° S. to Nova Scotia (according to Wilson, in Jardine's edition, vol. iii. p. 231,) in 45° N.; or exactly one hundred degrees of latitude. Its lesser range in Northern than in Southern America is probably due to the more excessive nature of the climate in the former hemisphere. It is said to be partly migratory during winter, in the Northern and even in the Middle States, and likewise on the shores of the Pacific. The *C. aura* is found in the extreme parts of Tierra del Fuego, and on the indented coast, covered with thick forests, of West Patagonia, (but not on the arid plains of Eastern Patagonia,) in Chile, where it is called Jote, in Peru, in the West Indies; and, according to Wilson, it remains even during winter, in New Jersey and Delaware, latitude 40°. It and one of the family of Polyborinæ are the only two carrion-feeding hawks, which have found their way to the Falkland Islands. The Turkey buzzard, as it is generally called by the English, may be recognized at a great distance from its lofty, soaring and most graceful flight. It is generally solitary, or, at most, sweeps over the country in pairs. In Tierra del Fuego, and on the west coast of Patagonia, it must live exclusively on what the sea throws up, and on dead seals: wherever these animals in herds were sleeping on the beach, there this vulture might be seen, patiently standing on some neighbouring rock. At the Falkland Islands it was tolerably common; but sometimes there would not be a single one near the settlement for several days together, and then many would suddenly appear. They were usually shy; a disposition which is remarkable, as being different from that of almost every other bird in this Archipelago. May we infer from this that they are migratory, like those of the northern hemisphere? In a female specimen killed there, the skin of the head was intermediate in colour between "scarlet and cochineal red,"* and the iris dark-coloured. D'Orbigny describes the iris as being bright scarlet; whilst Azara says it is "jaune léger." Is this difference owing to the sex and age, as certainly is the case with the condors? As a considerable degree of confusion has prevailed in the synonyms of this and the foregoing species, caused apparently by a doubt to which of them Molina applied the name of *Jote*, I would wish to call attention to the fact, that at the present time the *C. aura* in Chile goes by the name of *Jote*. Moreover, I think Molina's description by itself might have decided the question; he says, the head of the *Vultur jota* is naked, and covered only with a wrinkled and *reddish* (roxiza) skin.

FAMILY—FALCONIDÆ.

SUB-FAM. POLYBORINÆ, *Swains.*
(Caracaridæ, D'Orbigny.)

POLYBORUS BRASILIENSIS. *Swains.*

Polyborus vulgaris, *Vieillot.*
Falco Brasiliensis Auctorum; Caracara of Azara; Tharu of Molina; and Carrancha of the inhabitants of La Plata.

THIS is one of the commonest birds in South America, and has a wide geographical range. It is found in Mexico and in the West Indies. It is also, according to M. Audubon, an occasional visitant to the Floridas; it takes its name from Brazil, but is no where so common as on the grassy savannahs of La Plata. It generally follows man, but is sometimes found even on the most desert plains of Patagonia: in the northern part of that region, numbers constantly attended the line of road between the Rio Negro and the Colorado, to devour the carcasses of the animals which chanced to perish from fatigue. Although abundant on the open plains of this eastern portion of the continent, and likewise on the rocky and barren shores of the Pacific, nevertheless it inhabits the borders of the damp and impervious forests of Tierra del Fuego and of the broken coast of West Patagonia, even as far south as Cape Horn. The Carranchas (as the *Polyborus Brasiliensis* is called in La Plata) together with the *P. chimango*†, attend in great numbers the estancias and slaughtering houses in the neighbourhood of the Plata. If an

* In this work, whenever the particular name of any colour is given, or it is placed within commas, it implies, that it is taken from comparison with Patrick Syme's edition of Werner's Nomenclature of Colours.
† *Milvago Chimango* of this work.

animal dies in the plain, the *Cathartes atratus* or Gallinazo commences the feast, and then these two carrion-feeding hawks pick the bones clean. Although belonging to closely allied genera, and thus commonly feeding together, they are far from being friends. When the Carrancha is quietly seated on the branch of a tree, or on the ground, the Chimango often continues flying backwards and forwards for a long time, up and down in a semicircle, trying each time, at the bottom of the curve, to strike its larger relative. The Carrancha takes little notice, except by bobbing its head. Although the Carranchas frequently assemble in numbers, they are not gregarious; for in desert places they may be seen solitary, or more commonly by pairs. Besides the carrion of large animals, these birds frequent the borders of streams and the sea-beach, for the sake of picking up whatever the waters may cast on shore. In Tierra del Fuego, and on the west coast of Patagonia, they must live almost exclusively on this last means of supply.

The Carranchas are said to be very crafty, and to steal great numbers of eggs; they attempt also, together with the Chimango, to pick the scabs off the sore backs of both horses and mules. On the one hand, the poor animal, with its ears down and its back arched; and, on the other, the hovering bird, eyeing at the distance of a yard, the disgusting morsel, form a picture which has been described by Captain Head with his own peculiar spirit and accuracy. The Carranchas kill wounded animals; but Mr. Bynoe (the surgeon of the Beagle) saw one seize in the air a live partridge, which, however, escaped, and was for some time chased on the ground. I believe this circumstance is very unusual: at all events there is no doubt that the chief part of their sustenance is derived from carrion. A person will discover their *necrophagous* habits by walking out on one of the desolate plains, and there lying down to sleep: when he awakes, he will see on each surrounding hillock, one of these birds patiently watching him with an evil eye. It is a feature in the landscape of these countries, which will be recognised by every one who has wandered over them. If a party goes out hunting with dogs and horses, it will be accompanied during the day, by several of these attendants. The uncovered craw of the Carrancha, after feeding, protrudes from its breast; at such times it is, and indeed generally, an inactive, tame, and cowardly bird. Its flight is generally heavy and slow, like that of the English carrion crow, whose place it so well supplies in America. It seldom soars; but I have twice seen one at a great height gliding through the air with much ease. It runs (in contradistinction to hopping), but not quite so quickly as some of its congeners. At times the Carrancha is noisy, but is not generally so; its cry is loud, very harsh and peculiar, and may be compared to the sound of the Spanish guttural *g*, followed by a rough double *r r*. Perhaps the Spaniards of Buenos Ayres, from this cause, have called it Carrancha. Molina, who says it is called Tharu in Chile, states, that when uttering this cry, it elevates its head higher and higher, till at last, with its beak wide open, the crown almost touches the lower part of the back. This fact, which has been doubted, is true; for I have myself several times seen them with their heads backwards, in a completely inverted position. The Carrancha builds a large coarse nest, either in a low cliff, or in a bush or lofty tree. To these observations I may add, on the high authority of Azara, whose statements have lately been so fully confirmed by M. D'Orbigny, that the Carrancha feeds on worms, shells, slugs, grasshoppers, and frogs; that it destroys young lambs by tearing the umbilical cord: and that it pursues the Gallinazos and gulls which attend the slaughtering-houses, till these birds are compelled to vomit up any carrion they may have lately gorged. Lastly, Azara states that several Carranchas, five or six together, will unite in chase of large birds, even such as herons. All these facts show that it is a bird of very versatile habits and considerable ingenuity.

I am led to suppose that the young birds of this species sometimes congregate together. On the plains of Santa Cruz (lat. 50° S. in Patagonia), I saw in the month of April, or early autumn, between twenty and thirty Polybori, which I at first thought would form a species distinct from *P. Brasiliensis*. Amongst those I killed, there were some of both sexes; but the ovarium in the hens was only slightly granular. The plumage of the different individuals was nearly similar; and in none appeared like that of an adult bird, although certainly not of a very young one. Having mentioned these circumstances to Mr. Gould, he likewise suspected it would form a new species; but the differences appear so trifling between it and the specimens of young birds in the British Museum and in the Museum of the Zoological Society, and likewise of the figure of a young bird given by Spix, (Avium Species Novæ, vol. i. p. 3.), that I have thought it advisable merely to allude to the circumstance. In my specimen, which is a cock, the head, instead of being of a dark brown, which is the usual character of even very immature birds, is of a pale rusty brown. The bill and cere are less produced than in the adult *P. Brasiliensis;* and the cere is of a brighter colour, than what appears to be usual in the young of this species. In other respects there is such a perfect similarity between them, that I do not hesitate to consider my specimen as a young bird of the *P. Brasiliensis* in one of its states of change;—and to be subject to great variation of plumage during growth, is known to be a character common to the birds of this sub-family. It may, however, possibly be some variety of the *P. Brasiliensis*, for this bird seems subject to variation: Azara (Voyage dans l'Amérique Méridionale, vol. iii. p. 35.) remarks, "Il y a des individus dont les teintes sont plus faibles, ou d'un brun pâle, avec des taches sur la poitrine, et d'autres qui ont des couleurs plus foncées; j'ai décrit ceux qui tiennent le milieu entre les uns et les autres."

I have myself more than once observed a single very pale-coloured bird, in form like the *P. Brasiliensis*, mingled with the other carrion-feeding hawks on the banks of the Plata; and there is now in the British Museum a specimen, which may be considered as partly an albino. Spix, on the other hand, (Avium Species Novæ, p. 3.) has described some specimens from the coast of Brazil, as being remarkable from the darkness of the plumage of their wings.

MILVAGO, *Spix*.

Several new genera have lately been established to receive certain species of the sub-family of *Polyborinæ*, and consequently great confusion exists in their arrangement. Mr. George R. Gray has been kind enough to give me the following observations, by which it appears he has clearly made out, that Spix's genus *Milvago*, is that which ought to be retained. M. D'Orbigny has made two sections in the genus *Polyborus*, according as the craw is covered with feathers, or is naked, and he states that the *P. Brasiliensis* is the only species which comes within the latter division; but we shall afterwards see that the *Falco Novæ Zelandiæ*, Auct. (the *Milvago leucurus* of this work) has a naked craw, which is largely protruded after the bird has eaten. M. D'Orbigny has also instituted the genus *Phalcobænus*, to receive a bird of this sub-family, with the following characters:

"Bec fortement comprimé, sans aucune dent ni sinus, à commissure très-arquée à son extrémité; cire alongée et droite; un large espace nu entourant la partie antérieure et inférieure de l'œil, et s'étendant sur toute la mandibule inférieure; tarses emplumés sur un tiers de leur longueur, le reste réticulé; doigts longs, semblables à ceux des gallinacés, terminés par les ongles longs, deprimés et élargis, très-peu arqués, toujours à extrémité obtuse ou fortement usée; ailes de la famille, la troisième penne plus longue que les autres."

Mr. George R. Gray, however, has pointed out to me that Spix, (in his Avium Species Novæ) ten years since, made a division in this sub-family, from the rounded form of the nostril of one of the species, namely, the *M. ochrocephalus* of his work, or the *Chimachima* of Azara. And Mr. Gray thinks, that all the species may be grouped much more nearly in relation to their affinities by this character, than by any other: he further adds ;—" The only difference which I can discover between this latter genus (*Milvago*), and D'Orbigny's (*Phalcobænus*), is, that in the latter the bill is rather longer, and not quite so elevated in the culmen as in the former; and these characters must be considered too trivial for the foundation of a generic division. I, therefore, propose to retain Spix's genus, *Milvago*, for all those *Polyborinæ* which possess *rounded nostrils with an elevated bony tubercle in the centre.* They were once considered to form three distinct genera, viz.—Milvago, *Spix.* (Polyborus, *Vieill.* Haliaëtus, *Cuv.* Aquila, *Meyen.*) —Senex, *Gray.* (Circaëtus, *Less.*)—Phalcobænus, *D'Orb.* but a careful comparison of the several species, shows a regular gradation in structure from one to the other, which induces me to consider them as only forming two sections of one genus. Those which have the bill short, with the culmen arched, and are of small size, slender form, and with the tarsi rather long and slender, are—

1. Milvago ochrocephalus, *Spix.*
   Polyborus chimachima; *Vieill.* (young).
   Falco degener, *Licht.*
   Haliaëtus chimachima, *Less.*
2. Milvago pezoporos, *nob.*
   Aquila pezopora, *Meyen.*
3. Milvago chimango, *n.*
   Polyborus chimango, *Vieill.*
   Haliaëtus chimango, *Less.*

Those which have a buteo-like appearance, and with rather short and stout tarsi, are,

7. Milvago leucurus, *n.*
   Falco leucurus, *Forster's* Drawings No. 34.
   Falco Novæ Zealandiæ, *Gm.*
   —— Australis, *Lath.*
   Circaëtus antarcticus, *Less.*
8. Milvago albogularis, *n.*
   Polyborus (Phalcobænus ?) albogularis, *Gould.*
9. Milvago montanus, *n.*
   Phalcobænus montanus, *D'Orbig.*
10. Milvago megalopterus, *n.*
    Aquila megaloptera, *Meyen.*

1. MILVAGO PEZOPOROS.

Aquila pezopora, *Meyen.* Nov. Act. Phys. Med. Acad. Cæs. Leo. Car. Nat. Cur. suppl. 1834. p. 62. pl. VI.

I obtained two specimens of this bird, one from Port Desire, in Patagonia, and another at the extreme southern point of Tierra del Fuego. Meyen* describes it as common on the plains of Chile, and on the mountains to an elevation of 4000 or 5000 feet. As M. D'Orbigny does not notice this species, I presume it is not found on the Atlantic side of the continent, so far north as the Rio Negro, where he resided for some time. The habits and general appearance of *M. chimango* and this bird are so entirely similar, that

* Novorum Actorum Academiæ Cæsariæ, Leopol. vol. xvi. p. 62. Observationes Zoologicas, F. J. Meyenii.

I did not perceive that the species were different; hence I cannot speak with certainty of their range, but it would appear probable that the *M. pezoporus* replaces in Chile, Tierra del Fuego and Southern Patagonia the *M. chimango* of La Plata. In the same manner the *M. chimango* is replaced between the latitudes of Buenos Ayres and Corrientes by a third closely allied species, the *M. ochrocephalus*. D'Orbigny, (p. 614, in the Zoological part of his work) speaking of the Chimango, says, "Il n'est pas étonnant qu'on ait long-temps confondu cette espèce avec le *falco degener*, Illiger, (the *M. ochrocephalus*) et qu'on l'ait cru de sa famille. Il est impossible de présenter plus de rapports de forme et surtout de couleur. Nous les avions, nous-même confondus au premier abord; mais, en remarquant, ultérieurement, que le sujet que nous regardions comme le mâle ne se trouvait qu'à Corrientes, tandis qu'il y avait seulement des femelles sur les rives de la Plata, l'étude plus attentive des mœurs de ces oiseaux, et les localités respectives qu'habite chacun d'eux, ne tarda pas à nous y faire reconnaître, avec Azara, deux espèces vraiment très-distinctes; mais qui, depuis, ont encore été confondues, sous la même nom, par M. la Prince Maximilien de Neuwied. *" I may observe that the figure given in Meyen's work, has the iris coloured bright red, instead of which it should have been brown.

2. MILVAGO CHIMANGO.

Polyborus chimango, *Vieill.*
Haliaëtus chimango, *Less.*
Chimango, *Azar.* Voyage, vol. iii. p. 35.

My specimen was obtained at Maldonado, on the banks of the Plata. In the following short account of the habits of this bird, it must be understood that I have confounded together, the *M. chimango* and the *M. pezoporus;* but I am certain that almost every remark is applicable to both species. From what has been said under the last head, it may be inferred, that both of these allied birds have comparatively limited ranges, compared with that of the *P. Brasiliensis*. Azara says the Chimango (and he first distinguished this species from the *M. ochrocephalus*, or *M. chimachima*) is rarely found so far north as Paraguay. D'Orbigny saw the Chimango (*M. pezoporus ?*) at Arica in lat. 16°, and I killed the *M. pezoporus* in the extreme southern point of America, in lat. 55° 30' south.

The Chimango, in La Plata, lives chiefly on carrion, and generally is the last bird of its tribe which leaves the skeleton, and hence it may frequently be seen standing within the ribs of a cow or horse, like a bird in a cage. The Chimango often frequents the sea-coast and the borders of lakes and swamps, where it picks up small fish. It is truly omnivorous, and will eat even bread, when thrown out

* Tom. iii. p. 162.

of a house with other offal. I was also assured that in Chiloe, these birds (probably in this district the *M. pezoporus*) materially injure the potato crops, by stocking up the roots when first planted. In the same island, I saw them following by scores the plough, and feeding on worms and larvæ of insects. I do not believe that they kill, under any circumstances, even small birds or animals. They are more active than the Carranchas, but their flight is heavy; I never saw one soar; they are very tame; are not gregarious; commonly perch on stone walls, and not upon trees. They frequently utter a gentle, shrill scream.

3. Milvago leucurus.

Falco leucurus, *Forster's* Drawings, No. 34. MS.
—— Novæ Zelandiæ, *Gm.*
—— australis, *Lath.*
Circaëtus antarcticus, *Less.*

It will be observed in the above list of synonyms, which I have given on the authority of Mr. G. R. Gray, that this bird, although possessing well marked characters, has received several specific names. Mr. Gray's discovery of Forster's original drawing with the name *F. leucurus* written on it, I consider very fortunate, as it was indispensable that the names by which it is mentioned in most ornithological works, namely, *Falco* or *Polyborus Novæ Zelandiæ*, should be changed. There is not, I believe, the slightest reason for supposing that this bird has ever been found in New Zealand. All the specimens which of late years have been brought to England have come from the Falkland Islands, or the extreme southern portion of South America. The sub-family, moreover, to which it belongs, is exclusively American; and I do not know of any case of a land bird being common to this continent and New Zealand. The origin of this specific name, which is so singularly inappropriate, as tending to perpetuate a belief which would form a strange anomaly in the geographical distribution of these birds, may be explained by the circumstance of specimens having been first brought to Europe by the naturalists during Captain Cook's second voyage, during which New Zealand was visited, and a large collection made there. In the homeward voyage, however, Cook anchored in Christmas Sound, in Tierra del Fuego, and likewise in Staten Land: describing the latter place he says, " I have often observed the *eagles* and *vultures* sitting on the hillocks among the shags, without the latter, either young or old, being disturbed at their presence. It may be asked how these birds of prey live? I suppose on the carcasses of seals and birds, which die by various causes; and probably not few, as they are so numerous." From this description I entertain very little doubt that Cook referred to the *Cathartes aura* and *Milvago leucurus*, both of which birds inhabit these latitudes, as we shall hereafter show.

The plumage in the two sexes of this species differs in a manner unusual in the family to which it belongs. The description given in all systematic works is applicable, as I ascertained by dissection, only to the old females; namely, back and breast black, with the feathers of the neck having a white central mark following the shaft,—tectrices, with a broad white band at extremity; thighs and part of the belly rufous-red; beak "ash gray," with cere and tarsi "Dutch orange."

MALE of smaller size than female: dark brown; with tail, pointed feathers of shoulders and base of primaries, pale rusty brown. On the breast, that part of each feather which is nearly white in the female, is pale brown: bill black, cere white, tarsi gray. As may be inferred from this description, the female is a much more beautiful bird than the male, and all the tints, both of the dark and pale colours, are much more strongly pronounced. From this circumstance, it was long before I would believe that the sexes were as here described. But the Spaniards, who are employed in hunting wild cattle, and who (like the aboriginal inhabitants of every country) are excellent practical observers, constantly assured me that the small birds with gray legs were the males of the larger ones with legs and cere of an orange colour, and thighs with rufous plumage.

The YOUNG MALE can only be distinguished from the adult bird by its beak not being so black, or cere so white; and likewise in a trifling difference of plumage, such as in the markings of the pointed feathers about the head and neck, being more like those of the female than of the old cock. One specimen, which I obtained at the Falkland Islands, I suppose is a one-year-old female; but its organs of generation were smooth: in size larger than the male; the tail dark brown, with the tip of each feather pale colour, instead of being almost black with a white band; under tail-coverts dark brown, instead of rufous; thighs only partly rufous, and chiefly on the inner sides; feathers on breast and shoulder like those of male, with part near shaft brown; those on back of head with white, like those of adult females. Beak, lower mandible gray, upper black and gray (in the old female the whole is pale gray); the edge of cere and the soles of the feet orange, instead of the whole of the cere, tarsi, and toes being thus coloured. The circumstance of the young birds of, at least, one year and a half old, as well as of the adult males, being brown coloured, will, I believe, alone account for the singular fewness of the individuals with rufous thighs, a fact which at first much surprised me.

The *Milvago leucurus* is exceedingly numerous at the Falkland Islands, and, as an old sealer who had long frequented these seas remarked to me, this Archipelago appears to be their metropolis. I was informed, by the same authority, that they are found on the Diego Ramirez Rocks, the Il Defonso islands, and on some others, but never on the mainland of Tierra del Fuego. This statement I can corroborate to a certain degree, since I never saw one in the southern part of

Tierra del Fuego, near Cape Horn, which was twice visited during our voyage. They are not found on Georgia, or on the other antarctic islands. In many respects these hawks very closely resemble in their habits the *P. Brasiliensis*. They live on the flesh of dead animals, and on marine productions. On the Ramirez Rocks, which support no vegetation, and therefore no land-animals, their entire sustenance must depend upon the sea. At the Falkland Islands they were extraordinarily tame and fearless; and constantly haunted the neighbourhood of the houses to pick up all kinds of offal. If a hunting party in the country killed a beast, these birds immediately congregated from all quarters of the horizon; and standing on the ground in a circle, they patiently awaited for their feast to commence. After eating, their uncovered craws are *largely* protruded, giving to them a disgusting appearance. I mention this particularly, because M. D'Orbigny says that the *P. Brasiliensis* is the only bird of this family in which the craw is much developed. They readily attack wounded birds; one of the officers of the Beagle told me he saw a cormorant in this state fly to the shore, where several of these hawks immediately seized upon it, and hastened its death by their repeated blows. I have been told that several have been seen to wait together at the mouth of a rabbit hole, and seize on the animal as it comes out. This is acting on a principle of union, which is sufficiently remarkable in birds of prey; but which is in strict conformity with the fact stated by Azara, namely, that several Carranchas unite together in pursuit of large birds, even such as herons.

The Beagle was at the Falkland Islands only during the early autumn (March), but the officers of the Adventure, who were there in the winter, mentioned many extraordinary instances of the boldness and rapacity of these birds. The sportsmen had difficulty in preventing the wounded geese from being seized before their eyes; and often, when having cautiously looked round, they thought they had succeeded in hiding a fine bird in some crevice of the rocks, on their return, they found, when intending to pick up their game, nothing but feathers. One of these hawks pounced on a dog which was lying asleep close by a party, who were out shooting; and they repeatedly flew on board the vessel lying in the harbour, so that it was necessary to keep a good look-out to prevent the hide used about the ropes, being torn from the rigging, and the meat or game from the stern. They are very mischievous and inquisitive; and they will pick up almost anything from the ground: a large black glazed hat was carried nearly a mile, as was a pair of heavy balls, used in catching wild cattle. Mr. Usborne experienced, during the survey, a severe loss, in a small Kater's-compass, in a red morocco case, which was never recovered. These birds are, moreover quarrelsome, and extremely passionate; it was curious to behold them when, impatient, tearing up the grass with their bills from rage. They are not truly gregarious; they do not soar, and their flight is heavy and clumsy. On the ground they run with extreme quickness, putting out one leg before the other, and stretching forward their bodies, very much like pheasants. The sealers, who have sometimes, when pressed by hunger, eaten them, say that the flesh when cooked is quite white, like that of a fowl, and very good to eat—a fact which I, as well as some others of a party from the Beagle, who, owing to a gale of wind, were left on shore in northern Patagonia, until we were very hungry, can answer for, is far from being the case with the flesh of the Carrancha, or *Polyborus Brasiliensis*. It is a strange anomaly that any of the *Falconidæ* should possess such perfect powers of running as is the case with this bird, and likewise with the *Phalcobænus montanus* of D'Orbigny. It perhaps, indicates an obscure relationship with the Gallinaceous order—a relation which M. D'Orbigny suggests is still more plainly shown in the Secretary Bird, which he believes represents in Southern Africa, the *Polyborinæ* of America.

The *M. leucurus* is a noisy bird, and utters several harsh cries; of which, one is so like that of the English rook, that the sealers always call it by this name. It is a curious circumstance, as shewing how, in allied species, small details of habit accompany similar structure, that these hawks throw their heads upwards and backwards, in the same strange manner, as the Carranchas (the Tharu of Molina) have been described to do. The *M. leucurus*, builds on the rocky cliffs of the sea-coast, but (as I was informed) only on the small outlying islets, and never on the two main islands: this is an odd precaution for so fearless a bird.

4. MILVAGO ALBOGULARIS.

PLATE I.

Polyborus, (Phalcobænus) albogularis, *Gould*, Proceedings of Zoolog. Soc. Part V. (Jan. 1837.) p. 9.

M. Fœm. fuscescenti-niger, marginibus plumarum inter scapulas fulvis; primariis secundariisque albo ad apicem notatis; gulâ, pectore, corporeque subtùs albis; lateribus fusco sparsis; rostro livido, lineis nigris ornato; cera tarsisque flavis.

LONG. tot. 20 unc. ½; rostri, 1⅜; alæ, 15¾; caudæ, 9; tarsi, 3.

Description of female specimen, believed to be applicable to both sexes.

COLOUR.—Head, back, upper wing coverts pitch black, passing into liver brown; feathers on back of neck and shoulders terminating in a yellowish-brown tip, of which tint the external portion of the primaries, and nearly the whole of the tertiaries partake. Tail liver brown, with a terminal white band nearly one inch broad; base of the tectrices white, irregularly marked with brown: upper tail coverts white. All the feathers of the wing tipped with white, their bases irregularly barred with transverse marks of brown and white. *Under surface.*—Chin, throat, breast, belly, thighs, under tail-coverts, under lining of wings, and edge of shoulders perfectly white. On the flanks, however, there are some brown feathers irregularly interspersed; and on the lower part of the breast, most of the feathers show a most obscure margin of pale brown. Bill horn-colour. Cere and tarsi yellow.

FORM.—Cere and nostril as in the *M. Leucurus*, but the bill not quite so strong. Feathers on the sides and back of head narrow and rather stiff; those on the shoulders obtusely pointed,—which character of plumage is very general in this sub-family. Wing: fourth primary very little longer than the third or the fifth, which are equal to each other. First primary three inches shorter than the fourth or longest, and more nearly equal to the sixth than to the seventh. Extremity of wing reaching to within about an inch and a half of the tail. Tarsi reticulated, with four large scales at the base: upper part covered with plumose feathers for about three quarters of an inch below the knee; but these feathers hang down and cover nearly half of the leg. Middle toe with fifteen scales, outer ones with about nine. Claws of nearly the same degree of strength, curvature and breadth as in *Polyborus Brasiliensis*, or in *M. leucurus*, but sharper than those of the latter.

	Inch.		Inch.
Total length	20¼	Hind claw measured in straight line from tip to root	⅚
Tail	9		
Wings when folded	15¾	Claw of middle toe, a twentieth less than that of the hind one.	
From tip of beak to anterior edge of eye	1⅝		
Tarsus from soles of feet to knee joint	3½		

Habitat, Santa Cruz, 50° S. Patagonia. (*April.*)

Mr. Gould, at the time of describing this species, entertained some doubts whether it might not eventually prove to be the *Phalcobænus montanus* of D'Orbigny, in a state of change. I have carefully compared it with the description of the *P. montanus*, and certainly, with the exception of the one great difference of *M. albogularis* having a white breast, whilst that part in the *P. montanus* is black, the points of resemblance are numerous and exceedingly close. The *M. albogularis*, appears to be rather larger, and the proportional length of the wing feathers are slightly different; the cere and tarsi are not of so bright a colour; the middle toe has fifteen scales on it instead of having sixteen or seventeen. The black shades of the upper surface are pitchy, instead of having an obscure metallic gloss, and the feathers of the shoulders are terminated with brown, so as to form a collar, which is not represented in the figure of

P. montanus, given by M. D'Orbigny. Although the main difference between the two birds, is the colour of their breasts, yet it must be observed, that in the *M. albogularis* there is some indication of an incipient change from white to brown in the plumage of that part. But as M. D'Orbigny, who was acquainted with the young birds of the *P. montanus,* (of which he has given a figure), does not mention so remarkable a modification in its plumage, as must take place on the supposition of *M. albogularis* being an immature bird of that species; and as the geographical range of the two is so very different, I am induced to consider them distinct. Moreover, on the plains of Santa Cruz, I saw several birds, and they appeared to me similar in their colouring. The *M. albogularis* is remarkable from the confined locality which it appears to frequent. A few pair were seen during the ascent of the river Santa Cruz, (Lat. 50° S.) to the Cordillera; but not one individual was observed in any other part of Patagonia. They appeared to me to resemble, in their gait and manner of flight, the *P. Brasiliensis;* but they were rather wilder. They lived in pairs, and generally were near the river. One day I observed a couple standing with the Carranchas and *M. pezoporus,* at a short distance from the carcass of a guanaco, on which the condors had commenced an attack. These peculiarities of habit are described by M. D'Orbigny in almost the same words, as occurring with the *P. montanus;* both birds frequent desert countries; the *P. montanus,* however, haunts the great mountains of Bolivia, and this species, the open plains of Patagonia.

In the valleys north of 30° in Chile, I saw several pair, either of this species, or of the *P. montanus* of D'Orbigny, (if, as is probable, they are different) or of some third kind. From the circumstance of its not extending (as I believe) so far south even as the valley of Coquimbo, it is extremely improbable that it should be the *M. albogularis,*—an inhabitant of a plain country twenty degrees further south. On the other hand, the *P. montanus* lives at a great elevation on the mountains of Upper Peru; and therefore it is probable that it might be found in a higher latitude, but at a less elevation. M. D'Orbigny says, "Elle aime les terrains secs et dépourvus de grands végétaux, qui lui seraient inutiles; car il nous est prouvé qu'elle ne se perche pas sur les branches." In another part he adds, "Elle descend cependant quelquefois jusque près de la mer, sur la côte du Pérou, mais ce n'est que pour peu de temps, et peut-être afin d'y chercher momentanément une nourriture qui lui manque dans son séjour habituel; peut-être aussi la nature du sol l'y attire-t-elle; car elle y trouve les terrains arides qui lui sont propres."* This is so entirely the character of the northern parts of Chile, that, it appears to me extremely probable, that the *P. montanus,* which inhabits the great mountains of Bolivia, descends, in Northern Chile, to near the shores of the Pacific; but that further

* Voyage dans l'Amerique Meridionale Partie, Oiseaux, p. 52.

south, and on the opposite side of the Cordillera, it is replaced by an allied species,—the *M. albogularis* of Santa Cruz.

5. MILVAGO MEGALOPTERUS.

Aquila megaloptera, *Meyen*, Nov. Act. Acad. Cæs. Suppl. 1834, p. 64. Pl. VIII.

When ascending the Despoblado, a branch of the valley of Copiapó in Northern Chile, I saw several brown-coloured hawks, which at the time appeared new to me, but of which I did not procure a specimen. These I have no doubt were the *A. megaloptera* of Meyen. In the British Museum there is a specimen, brought from Chile by Mr. Crawley. Mr. G. R. Gray suspects that this bird may eventually prove to be the young of the *Phalcobœnus montanus* of D'Orbigny, and as I saw that bird (or another species having a close general resemblance with it) in the valleys of Northern Chile, although not in the immediate vicinity, this supposition is by no means improbable. Meyen's figure at first sight appears very different from that of the young of the *P. montanus*, given by M. D'Orbigny, for in the latter the feathers over nearly the whole body are more distinctly bordered with a pale rufous shade, the thighs barred with the same, and the general tint is of a much redder brown. But with the exception of these differences, which are only in degree, I can find in M. D'Orbigny's description no other distinguishing character, whilst on the other hand, there are numerous points of close resemblance between the two birds in the shadings, and even trifling marks of their plumage. Meyen, moreover, in describing the habits of his species, says, it frequents a region just below the limit of perpetual snow, and that it sometimes soars at a great height like a condor. Those which I saw had the general manners of a *Polyborus* or *Milvago*, and were flying from rock to rock amongst the mountains at a considerable elevation, but far below the snow-line. In these several respects, there is a close agreement with the habits of the *P. montanus*, as described by M. D'Orbigny. I will only add that the specimen in the British Museum appeared, independently of differences of plumage, distinct from the *M. albogularis* of Patagonia, from the thinness and greater prolongation of its beak, and the slenderness of its tarsi.

Sub.-Fam.—BUTEONINÆ.
Craxirex. *Gould.*

Rostrum Buteonis sed longius; mandibulæ superioris margo rectus; versus apicem subitò incurvus. Alæ elongatæ. Cera lata. Nares ferè rotundæ, apertæ. Tarsi mediocres, anticè squamis tecti. Digiti magni, fortes; ungues obtusæ.

Mr. Gould was partly led to institute this genus from the facts communicated to him by me regarding the habits of the following species, which is found in the Galapagos Archipelago, and there supplies the place of the Polybori and Milvagines of the neighbouring continent of America. If a principle of classification founded on habits alone, were admissible, this bird, as will presently be shown, undoubtedly would be ranked with more propriety in the sub-family of Polyborinæ, than amongst the Buzzards. To the latter it is closely related in the form of its nostrils; in the kind of plumage which covers the head, breast, and shoulders; in the reticulation of the scales on its feet and tarsi, and less closely in the form of its beak. To the Polyborinæ it manifests an affinity in the great strength and length of its toes and claws, and in the bluntness of the latter; in the nakedness of the cere, in the perfectly uncovered nostrils, in the prolongation and bulk of the bill, in the straightness of the line of commissure, and in the narrow shape of the head. In these several respects, taken conjointly with its habits, this bird supplies a most interesting link in the chain of affinities, by which the true buzzards pass into the great American sub-family of carrion-feeding hawks. I am, indeed, unable to decide, whether I have judged rightly in placing this genus, as first of the Buteoninæ, instead of last of the Polyborinæ.

CRAXIREX GALAPAGOENSIS. *Gould.*
PLATE II.

Polyborus Galapagoensis. Proceedings of the Zoological Society for January, 1837, p. 9.

C. Mas. adult. Intensè fuscus; primariis nigris; secundariarum pogoniis internis transversim albo et fusco striatis; caudâ cinerascenti-fuscâ, transversim lineis angustis et numerosis intensè fuscis notatâ; rostro obscure corneo; pedibus olivaceo-flavis.

Long. tot. 20¼ unc.; *rostri*, 1½; *alæ*, 15; *caudæ*, 8½; *tarsi*, 3¼.

Fæm. adult. fæminæ juniori ferè similis, pectore tamen fusco.

Fæm. juv. Capite corporeque intensè stramineis, fusco-variegatis; illo in pectore et abdomine prævalente; primariis fusco-nigris; rectricum pogoniis externè cinerascenti-fuscis, internè pallide rosaceis; utrisque lineis angustis et frequentibus fuscis transversim striatis, apicibus sordide albis; rostro nigrescenti-fusco; pedibus olivaceo-flavis.

Long. tot. 24 unc.; *rostri*, 1¾; *alæ*, 17¼; *caudæ*, 10½; *tarsi*, 3½.

Description of adult male.

COLOUR.—Entire dorsal aspect umber brown: base of feathers on hind part of neck, white; base of those on back, irregularly banded with pale fulvous, and the scapulars with a distinct band of it. The inferior feathers of upper tail coverts banded in like manner to their extremities. Tail dusky clove-brown, obscurely marked with darkened transverse narrow bands. Primaries perfectly black towards their extremities, but with the outer edge of their base, gray: inner web banded and freckled with gray, brown, and white, which in the secondaries takes the form of regular bars. *Under surface*, entirely umber brown, but rather paler than the upper. Lining of wings gray, with irregular transverse brown bars: under-side of tail the same, but paler. Thighs of a rather yellower brown. Bill and cere horn colour, mottled with pale gray: tarsi yellow.

FORM.—Beak, with apex much arched, both longer and more pointed than it is in the group of the Polyborinæ. Cere naked, with few bristles; nostrils large, quite uncovered, irregularly triangular, with the angles much rounded, and situated rather above a central line between the culmen and commissure. Fourth primary longest, but third and fifth nearly equal to it; first, four inches and a half shorter than fourth, and equal to the eighth; second shorter than fifth. Extremities of wing reaching within half an inch of end of tail.

Tarsi strong, feathered for nearly a third of their length beneath the joint. Scales in narrow, undivided (with the exception in some instances of one) bands, covering the front of tarsus. Toes very strong and rather long, like those of the species of *Milvago*, and much more so than in the genus *Buteo*. Hind-toe equal in length to the inner one; but not placed quite so high on the Tarsus as in *Polyborus*. Basal joints of middle toe covered with small scales, with five large ones towards the extremity. Claws very strong, thick and long, and rather more arched, and broader than in *Polyborus Brasiliensis;* their extremities obtuse, but not in so great a degree as in some species of *Milvago*.

	Inches.
Total length from tip of bill to end of tail following curvature of body	20¼
Tail	8½
Wing, from elbow-joint to extremity of longest primary	15
Bill, from tip to anterior edge of eye measured in a straight line	$\frac{7}{10}$
Tarsus, from soles of feet to centre of joint	3⅓
Hind claw from tip to root, measured in straight line	1$\frac{1}{10}$
Claw of middle toe	$\frac{95}{100}$

Old female.

COLOUR.—Nearly as in young female, but with the breast dark brown.

Young female.

COLOUR.—Head, back of neck, back, wing coverts and tertiaries barred and mottled, both with pale umber brown (of the same tint as in the male bird) and with pale fulvous orange. On head and back of neck, each feather is of the latter colour, with a mere patch of the brown on its tip; but in the longer feathers, as in the scapulars, upper tail coverts, inner web and part of outer of the tertiaries, each is distinctly barred with the dark brown. Tail as in the old male. Primaries black as in male, with the inner webs nearly white, and marked with short transverse bars. Under surface and thighs of the same fulvous orange, but some of the feathers, especially those on the breast, are marked with small spots of umber brown on their tips. Some of the longer feathers on the flanks, on the under tail coverts, and on the linings of the wing, have irregular bars of the same.

FORM AND SIZE.—Larger and more robust than the male. Total length 24 inches. Tail ten and a half inches long, and therefore longer in proportion to the wings than in the other sex. Wings from joint to end of primaries, 17¼.

Habitat, Galapagos Archipelago, (*October*).

BIRDS.   25

This bird is, I believe, confined to the Galapagos Archipelago, where on all the islands, it is excessively numerous. It inhabits, indifferently, either the dry sterile region near the coast, which, perhaps, is its most general resort, or the damp and wooded summits of the volcanic hills. This bird, in most of its habits and disposition, resembles the *Milvago leucurus*, or the *Falco Novæ Zelandiæ* of older authors. It is extremely tame, and frequents the neighbourhood of any building inhabited by man. When a tortoise is killed even in the midst of the woods, these birds immediately congregate in great numbers, and remain either seated on the ground, or on the branches of the stunted trees, patiently waiting to devour the intestines, and to pick the carapace clean, after the meat has been cut away. These birds will eat all kinds of offal thrown from the houses, and dead fish and marine productions cast up by the sea. They are said to kill young doves, and even chickens; and are very destructive to the little tortoises, as soon as they break through the shell. In these respects this bird shows its alliance with the buzzards. Its flight is neither elegant nor swift. On the ground it is able, like the *M. leucurus* and *Phalcobænus montanus* of D'Orbigny, to run very quickly. This habit which, as before observed, is so anomalous in the Falcons, manifests in a very striking manner the relation of this new genus with the *Polyborinæ*. It is, also, a noisy bird, and utters many different cries, one of which was so very like the shrill gentle scream of the *M. chimango*, that the officers of the "Beagle" generally called it either by this name, or from its larger size by that of *Carrancha*,—both names, however, plainly indicating its close and evident relationship with the birds of that family. The craw is feathered; and does not, I believe, protrude like that of the *P. Brasiliensis* or *M. leucurus*. It builds in trees, and the female was just beginning to lay in October. The bird of which the full figure has been given, is a young female, but of, at least, one year old. The old male-bird is of a uniform dusky plumage, and is seen behind. The adult female resembles the young of the same sex, but the breast is dark brown like that of the male. In precisely the same manner as was remarked in the case of the *M. leucurus*, these old females are present in singularly few proportional numbers. One day at James' Island, out of thirty birds, which I counted standing within a hundred yards of the tents, under which we were bivouacked, there was not a single one with the dark brown breast. From this circumstance I am led to conclude that the females of this species (as with the *M. leucurus*) acquire their full plumage late in life.

1. BUTEO ERYTHRONOTUS.

Haliaëtus erythronotus, *King*, in Zoological Journal, vol. iii. p. 424.
Buteo tricolor, *D'Orbigny*.

I obtained specimens of this bird from Chiloe and the Falkland Islands, and Captain King who first described it, procured his specimens from Port Famine, Lat. 53° 38' in Tierra del Fuego. M. D'Orbigny states that it has a wide range over the provinces of La Plata, central Chile, and even Bolivia; but in this latter country, it occurs only on the mountains, at an elevation of about 12,000 feet above the sea. The same author states, that it usually frequents open and dry countries; but as we now see that it is found in the dense and humid forests of Chiloe and Tierra del Fuego, this remark is not applicable. At the Falkland Islands, it preys chiefly on the rabbits, which have run wild and abound over certain parts of the island. This bird was considered by Captain King as a *Haliaëtus*; but Mr. Gould thinks it is more properly placed with the Buzzards. Captain King gave it the appropriate specific name of *erythronotus*, and, therefore, as Mr. Gould observes, the more recent one of *tricolor*, given by M. D'Orbigny, must be passed over.

2. BUTEO VARIUS. *Gould.*

Buteo varius, *Gould*, Proceedings of the Zoological Society, Part v. 1837, p. 10.

B. vertice corporeque supra intensè fuscis, plumis fulvo marginatis vel guttatis; primariis secundariisque cinereis, lineis numerosis fuscis transversim striatis; caudâ cinereâ, lineis angustis numerosis fuscis transversim notatâ; singulis plumis flavescenti-albo ad apicem notatis; gulâ fuliginosâ; pectore fulvo, lineâ interruptâ nigrescente a gulâ tendente circumdato; abdomine imo lateribusque stramineo et rufescenti-fusco variegatis; femoribus crissoque stramineis lineis transversalibus anfractis rufescenti-fuscis ornatis; rostro nigro; cerâ tarsisque olivaceis.

Long. tot. 21½; *alæ*, 16½; *caudæ*, 10; *tarsi*, 3¾.

COLOUR.—Head and back of neck umber brown, with edges of the feathers fringed with fulvous, (or buff orange with some reddish orange) and their bases white. Shoulders brown, with the feathers more broadly edged. Back the same, with the basal part of the feathers fulvous, with transverse bars of the dark brown. Tail blueish gray, with numerous, narrow, transverse, faint black bars. Tail-coverts pale fulvous, with irregular bars of dark fulvous and brown. Wings: primaries blackish gray, obscurely barred; secondaries and tertiaries more plainly barred, and tipped with fulvous. Wing coverts, dark umber brown, largely tipped, and marked with large spots, almost forming bars, of pale fulvous. *Under surface.*—Chin black ; throat and breast ochre yellow, with a narrow dark brown line on the shafts of the feathers, which, in those on the sides of the throat and breast expands into a large oval spot. Feathers on belly reddish brown, fringed and marked at base with the ochre yellow. Lining of wings ochre yellow, with numerous transverse bars of dark brown. Under-side of tail, inner webs almost white, outer pale gray, with very obscure transverse bars. Thighs, ochre yellow, with numerous zigzag transverse bars of pale reddish brown. Bill pale blackish; iris brown; tarsi gamboge yellow.

Form.—Fourth primary very little longer than third, and about half an inch longer than fifth. First rather shorter than seventh, and longer than eighth. Wings when folded reaching within two inches of the extremity of the tail.

	Inches.
Total length	21½
Length of tail	10
Wings when folded	16½
From tip of beak to within anterior edge of nostril, measured in straight line	8.5/10
Tarsi from soles of feet to middle of knee joint	3¾
Middle toe, measured from basal joint to tip of claw	2⅓

Habitat, Strait of Magellan, (*February*,) and Port St. Julian in Southern Patagonia, (*January*.)

3. Buteo ventralis. *Gould.*

Buteo ventralis, *Gould,* Proceedings of the Zoological Society, Part v. 1837, p. 10.

B. vertice corporeque intensè nitide fuscis, plumis dorsalibus purpurascentibus ; primariis nigris ; caudâ fuscâ, lineis obscurioribus cancellatâ numerosis, ad apicem sordidè albâ ; gulâ abdomine medio crissoque stramineo-albis ; pectoris corporisque lateribus fasciâ abdominali femoribusque flavescenti-albis fusco notatis, notis in femoribus rufescentibus ; tarsis per mediam partem anticè plumosis, rostro nigro ; cerâ tarsisque flavis.

Long. tot. 23 unc.; *alæ*, 15½ ; *caudæ*, 9¼ ; *tarsi*, 3¼.

Colour.—Head, back of neck, back, and wing-coverts, umber brown. Feathers on sides of throat edged with fulvous; those on lower parts of back with their basal parts marked with large white spots, edged with fulvous, but which do not show, until the feathers are ruffled. Tail of the same dark brown as the back, with many bars of pale brown, and extreme points tipped with dirty white. Tail-coverts same brown, with the more lateral ones marked with white and fulvous. Wings: primaries black, with the inner and basal webs brownish; secondaries and tertiaries brown, with obscure traces of paler transverse bars. *Under surface.*—Chin almost white; throat and breast very pale ochre yellow, with narrow brown lines on the shaft of the feathers, which expand into large marks on the sides of the upper part of the breast, and into regular spots on those of the belly. Lining of wing white, with brown spots on the feathers near their tips, like on those of the belly. Thighs very pale ochre yellow, with transverse bars of pale brown, appearing like inverted wedge-formed marks, with the apex on the shafts. Under tail-coverts almost white; under side of tail pale gray, with darker gray bars on the inner side of shafts. Bill blueish black, with base of lower mandible and part of upper yellowish. Tarsi pale yellow.

FORM.—Fourth primary very little longer than either the third or fifth, which are equal. First nearly equal to the eighth. Extremity of wing when folded reaching within two inches and a half of the end of the tail.

	In.		In.
Total length	23	Tarsi	$3\frac{1}{2}$
Wing when folded	$15\frac{1}{2}$	Middle toe from joint to tip of claw	3
Tail	$9\frac{1}{2}$	From extremity of beak to within nostril	$\frac{9}{10}$

Habitat, Santa Cruz, Lat. 50° S. Patagonia, (*April*.)

Mr. Gould remarks that "this species has all the characters of a true *Buteo*, and will rank as one of the finest of this well defined group. In size it rather exceeds the Common Buzzard of Europe, which in its general style of colouring it somewhat resembles."

SUB-FAM.—FALCONINA, VIG.

FALCO FEMORALIS. *Temm.*

Falco femoralis, *Temm.* Pl. Col. 121 male; and 343 adult male.
——— *Spix*, Av. Sp. Nov. 1. p. 18.

This specimen was shot in a small valley on the plains of Patagonia, at Port Desire, in Lat. 47° 44'. It builds its nest in low bushes, and the female was sitting on the eggs in the beginning of January. Egg, 1·8 of an inch in longer diameter, and 1·4 in shorter; surface rough with white projecting points; colour nearly uniform dirty "wood brown," thickly freckled with rather a darker tint; general appearance, as if it had been rubbed in brown mud. M. D'Orbigny supposed that Latitude 34° was the southern limit of this species; we now find its range three hundred and thirty miles further southward. The same author states that this falcon prefers a dry open country with scattered bushes, which answers to the character of the valleys, in the plains near Port Desire.

Tinnunculus Sparverius. *Vieill.*

Falco sparverius, *Linn. et Auct.*

I obtained specimens both from North and South Patagonia (Rio Negro and Santa Cruz), and Captain King found it at Port Famine in Tierra del Fuego. I saw it at Lima in Peru; and Mr. Macleay (Zoological Journal, vol. iii.) sent specimens from Cuba. According to Wilson it is common in the United States, and Richardson says its northern range is about 54°. The *Tinnunculus* therefore, ranges throughout both Americas over more than 107 degrees of latitude, or 6420 geographical miles. It is the only bird, which I saw in South America, that hovered over one particular spot, in the same stationary manner, as the common English kestrel (*Falco tinnunculus*, Linn.) is so frequently observed to do.

Sub-Fam.—CIRCINÆ.

1. Circus megaspilus. *Gould.*

Circus megaspilus, *Gould*, in Proceedings of the Zoological Society, Part V. 1837, p. 10.

C. vertice corporeque supra intensè fuscis, lineâ stramineâ a naribus supra oculos ad occiput tendente; hoc rufescenti-fusco; primariis intensè fuscis ad basin cinereis, lineis nigris cancellatis; caudæ tectricibus albis; rectricibus intermediis cinereis, externis cinereo-stramineis, omnibus lineis latis fuscis transversim notatis, lineâ ultimâ latissimâ, apice sordidè stramineo; gulâ pectoreque stramineis, fusco variegatis; corpore subtus stramineo; plumis pectoris laterumque striâ centrali fuscâ notatis; rostro nigro; cerâ tarsisque flavis.

Long. tot. 22 unc.; *rostri*, 1¼; *alæ*, 17; *caudæ*, 10½; *tarsi*, 3⅛.

COLOUR.—Head, back of throat, whole back, and wing-coverts umber brown, of a nearly uniform tint, and not very dark. Front, over the nostrils, with few fulvous bristly feathers; over the eyes, extending backward, a pale almost pure white streak, which joins an irregular band, extending across the nape of the neck, from below ear to ear, of brown feathers, edged with pale fulvous, giving a streaked appearance to that part. The wing-coverts are just tipped with dirty white. Wings: primaries of the same brown as the back, the inner ones assuming a gray tinge; these, and the basal parts of the inner webs of all, are obscurely barred; secondaries and tertiaries of a paler brown than the interscapular region. Tail grayish brown, with five well-defined bars, about ¾ of an inch wide, of the same brown, as the rest of the upper surface; extremities tipped with very pale dirty brown. Tail-coverts; upper ones brown, and the under ones white, with small brown spots on the shaft towards their extremities. *Under surface.* — Chin, pale fulvous, or ochre yellow. Breast, belly, thighs and under tail-coverts the same; the feathers on the lower part of the breast and on the belly have a dark brown mark along the shaft, which widens but very little towards the extremity; the brown on those on the upper part of the breast and on the throat is broader, and some of the feathers are of a darker fulvous, and as the dark brown of the back encroaches on each side, this part is much darker than the rest of the under surface. Above this, and just beneath the chin, a kind of collar is formed from ear to ear, of short feathers of a more strongly pronounced fulvous tint, with a narrow brown streak on their shafts. Lining of wings, and flanks almost white, with transverse brown bars. Under side of tail pale gray passing into fulvous, with the terminal dark brown bars seen through. Bill, horn-coloured, with some white markings towards its base; tarsi bright yellow.

FORM.—Third primary rather longer than fourth, second equal to fifth; first more nearly equal to the sixth than to the seventh. Wings reaching within an inch of the end of the tail. Feathers on thighs depend but little below the knee.

	In.		In.
Total length	22	Tarsi	3½
Wings folded	17	Middle toe to end of claw	2¾
Tail	10½	From tip of bill to nearest part of cere	$\frac{14}{100}$

Habitat, Maldonado, La Plata, (*July.*)

This hawk was not uncommon on the grassy savannahs and hills in the neighbourhood of the Rio Plata. Mr. Gould remarks "that in size it fully equals the *Circus æruginosus* of Europe, which it doubtless represents in the countries it inhabits. This species has a remarkable specific character in the lanceolate and conspicuous stripes down its breast."

2. CIRCUS CINERIUS. *Vieill.*

Circus cinerius, *Vieill.* Ency. Meth.
Falco histrionicus, *Quoy and Gaim.* Voy. autour du monde, Plate 15.
Circus histrionicus, *Vigors*, Zoological Journal, vol. iii. p. 425, note.

My specimens were obtained at the Falkland Islands, and at Concepçion in Chile. M. D'Orbigny states that it is a wild bird; but at the Falkland Islands it

was, for one of its order, very tame. The same author gives a curious account of
its habits: in a different manner from other raptorial birds, when it has killed its
prey, it does not fly to a neighbouring tree, but devours it on the spot. It roosts on
the ground, either on the top of a sand hillock, or by the bank of a stream : it
sometimes walks, instead of hopping, and when doing so, it has some resemblance
in general habit to the *Milvago chimango*. It preys on small quadrupeds, mollus-
cous animals, and even insects; and I find in my notes, that I saw one in the
Falkland Islands, feeding on the carrion of a dead cow. Although in these
respects this *Circus* manifests some relation in its habits with the *Polyborinæ*,
yet it has the elegant and soaring flight, peculiar to its family; and in form it
does not depart from the typical structure. Mr. Gould remarks that "we see in
this elegant bird as perfect an analogue of the *Circus cyaneus* of Europe, as in the
preceding species of the *Circus æruginosus*."

FAMILY.—STRIGIDÆ.

SUB-FAM.—SURNINÆ.

ATHENE CUNICULARIA. *Bonap.*

Strix cunicularia, *Mol. Bonap.* Am. Orni. I. 68. pl. 7. f. 2.

This bird, from its numbers and the striking peculiarities of its habits has
been mentioned in the works of all travellers, who have crossed the Pampas. In
Banda Oriental it is its own workman, and excavates its burrow on any level
spot of sandy soil; but in the Pampas, or wherever the Bizcacha is found, it uses
those made by that animal. During the open day, but more especially in the
evening, these owls may be seen in every direction standing frequently by pairs
on the hillock near their habitation. If disturbed, they either enter the hole, or,
uttering a shrill harsh cry, move with a remarkably undulatory flight to a short
distance, and then turning round, steadily gaze at their pursuer. Occasionally in
the evening they may be heard hooting. I found in the stomachs of two which I
opened the remains of mice; and I saw a small snake killed and carried away by
one. It is said that reptiles are the common object of their prey during the day
time. Before I was aware, from the numbers of mice caught in my traps, how
vastly numerous the small rodents are in these open countries, I felt much surprise how such infinite numbers of owls could find sufficient means of support.
I never saw this bird south of the Rio Negro, (Lat. 41° S.) In North America
they frequent only the trans-Mississippian territories in the neighbourhood of the
Rocky Mountains. The account given by Say of their habits, agrees with what

may every day be observed in the Pampas; but in the northern hemisphere they inhabit the burrows of the Marmot or Prairie dog, instead of those of the Bizcacha; and it would appear that their food is chiefly derived from insects, instead of from small quadrupeds and reptiles. Mr. Gould says he has compared my specimens from La Plata and Chile, on opposite sides of the Cordillera, with those from Mexico and the Rocky Mountains of North America, and he cannot perceive the slightest specific difference between them.

Sub-Fam.—ULULINÆ.

1. Otus Galapagoensis. Gould.

Plate III.

Otus (Brachyotus) Galapagoensis, *Gould*, in Proceedings of the Zoological Society, Part V., 1837, p. 10.

O. fasciâ circa oculos fuliginosâ; strigâ superciliari, plumis nares tangentibus et circa angulum oris, gulâ et disci facialis margine, albis; vertice corporeque supra intensè stramineo fuscoque variegatis; primariis ad apicem intensè fuscis, ad basin stramineo fasciatis; corpore subtus stramineo, notis irregularibus fasciisque fuscis ornato; femoribus tarsisque plumosis rufescenti-stramineis; rostro unguibusque nigris.

Long. tot. 13½; *rostri*, 1; *alæ*, 11; *caudæ*, 6; *tarsi*, 2.

Colour.—Facial disc; plumose feathers immediately around the eyes, nearly black, tipped with glossy fulvous; those nearer the margin are white at their base, and only slightly tipped with a darker brown. Between the eyes a band of small fulvous feathers with a central streak of dark brown, passing backward, blends into the plumage of the nape. Back of head and throat streaked with fulvous and brown, the centre of each feather being brown, and its edge fulvous. Interscapular region and the feathers of the wing, coloured in the same manner, but the fulvous part is indented on each side of the shaft in the brown, giving an obscurely barred appearance to these feathers. Primaries brown, with large rounded marks of fulvous; those on the first feather being smaller, and almost white: wing-coverts brown, and but little mottled. Tail with transverse bars of the same brown and fulvous, the latter colour much clearer and stronger on the external feathers; in the central ones, the fulvous part includes irregular markings of the dark brown. *Under surface.*—Throat and breast, with center of each feather brown, edged with fulvous; the former colour being predominant. On the belly and under tail-coverts the brown coloured marks on the shafts are narrow, but they are united to narrow transverse bars, which form at the points of intersection marked something like arrow-heads. The fulvous tint is here predominant. Downy feathers on thighs same fulvous colour as rest of body. Bill black.

FORM.—Second primary scarcely perceptibly longer than the first, and fourth rather longer than first. Tarsi thickly clothed with short feathers to the root of the nails.

	In.		In.
Total length	13½	Tarsi	2
Wings	11	Middle toe to root of nail	1¼
Tail	6	From tip of beak to interior edge of nostril	·⁄₂

Habitat, James Island, Galapagos Archipelago, (*October*).

Mr. Gould informs me, that "this species has most of the essential characters of the common short-eared owl of Europe (*Strix brachyota*), but differs from it, and all the other members of the group, in its smaller size and darker colouring."

The lesser proportional size of the fulvous marks on the first primaries, and on the tail, and the peculiar transverse brown marks on the feathers of the belly, easily distinguish it from the common short-eared owl. The specimen described is a male bird.

2. OTUS PALUSTRIS. *Gould*.

Strix brachyota. *Lath*.

Specimens of this bird were obtained at the Falkland Islands, at Santa Cruz in Patagonia, and at Maldonado on the northern bank of the Plata. At the latter place it seemed to live in long grass, and took to flight readily in the day. At the Falkland Islands it harboured in a similar manner amongst low bushes. Mr. Gould says, "So closely do the specimens brought home by Mr. Darwin, resemble European individuals, that I can discover no specific difference, by which they may be distinguished."

We have, therefore, the same species occurring in lat. 52° S. on the coast of South America, and in the northern division of the continent, according to Richardson, even as far as the sixty-seventh degree of latitude. Jardine says it is found in the Orkney islands (lat. 59°), and in Siberia; and that he has received specimens of it from Canton. M. D'Orbigny says it is found in the Sandwich and Marianne islands in the Pacific Ocean, and at Bengal in India. This bird, therefore, may be considered as a true cosmopolite.

ULULA RUFIPES.

Strix rufipes, *King*, in Zoological Journal, Vol. iii. p. 426.

I obtained a specimen of this bird from a party of Fuegians in the extreme southern islands of Tierra del Fuego. Owls are not uncommon in this country, and as small birds are not plentiful, and the lesser rodents extremely scarce, it at first appears difficult to imagine on what they feed. The following fact, perhaps, explains the circumstance: Mr. Bynoe, the surgeon to the "Beagle," killed an owl in the Chonos Archipelago, where the nature of the country is very similar to that of Tierra del Fuego, and, on opening its stomach, he found it filled with the remains of large-sized crabs: I conclude, therefore, that these birds here likewise subsist chiefly on marine productions.

Sub.-Fam.—STRIGINÆ.

1. Strix flammea. *Linn.*

I obtained a specimen of a white owl from Bahia Blanca in Northern Patagonia, and Mr. Gould remarks concerning it, that he only retains the name of *S. flammea* provisionally, until all the white owls, from various countries, shall have been subjected to a careful examination. Mr. Gould suspects, that when this is effected, the South American white owl will prove to be specifically distinct from that of Europe.

2. Strix punctatissima. *G. R. Gray.*

Plate IV.

S. supra nigricans, flavo subnebulosa, minutè albo-punctatissima, maculâ albâ ad apicem plumæ, cujusvis; subtus fulva, fasciis interruptis nigricantibus; caudâ dorso concolore, nigricanti-fasciatâ, apice albâ; disco faciali castaneo-rufo nigricanti-nebuloso circumdato, pogoniis internis albis, scapis nigris; pedibus longis, infra genu plumosis; tarso reliquo digitisque subpilosis.

Long. tot. $13\frac{1}{2}$; *alæ,* $9\frac{1}{4}$; *caudæ,* $4\frac{1}{4}$; *tarsi,* $2\frac{1}{10}$.

Colour.—Head and feathers within facial disc, glossy ferruginous brown, those forming the margin of it, same coloured, with their tips dark brown. Back of head and throat smoky brown, mottled with numerous small white dots, on the tips of the feathers. Back and wing-coverts the same, with the white spots larger and purer. Wings: primaries, same dark brown, mottled with dull chesnut red; the tip of each, with the exception of the three first, is marked with a triangular white spot, of the same kind with those over the rest of the body, but larger. Tail, transversely barred with brown and reddish fulvous, and the extreme points mottled with white. *Under surface.* Breast, belly and lining of wings, fulvous, mottled with brown;—the feathers being transversely barred with narrow brown lines. Under side of tail, pale gray, with well defined transverse bars of a darker gray. Short downy feathers on tarsi, of a brighter fulvous than the rest of the under surface.

FORM.—Third primary rather longer than second; first equal to third. Wing, exceeding the tail in length by nearly one inch and a quarter. Short feathers on the tarsus, extending about one-third of its length, below the knee. Tarsi, elongated. Toes and lower part of tarsi, with few scattered brown hairs.

	In.		In.
Total length	$13\frac{1}{2}$	Tarsi	$2\frac{7}{10}$
Wing	$9\frac{1}{4}$	Tip of beak to rictus	$1\frac{1}{4}$
Tail	$4\frac{1}{4}$	Middle toe, from root of claw to base	$1\frac{1}{10}$

Habitat, James Island, Galapagos Archipelago, (*October.*)

I am indebted to Mr. G. R. Gray for the description of this species, which is deposited in the British Museum. Only one specimen was obtained during our visit to the Galapagos Archipelago; and this formed part of the collection made by the direction of Captain FitzRoy.

This owl is in every respect a true Strix; it is fully a third less than the common species of Europe, and differs from it in many respects, especially in the darker colouring of its plumage. The colouring of the Plate is not perfectly accurate in its minuter details.

Family.—CAPRIMULGIDÆ.

Sub-Fam.—CAPRIMULGINÆ.

1. Caprimulgus bifasciatus. *Gould.*

Caprimulgus bifasciatus, *Gould*, in Proceedings of the Zoological Society, February 1837, p. 22.

C. capite nigro fusco et fulvescente ornatus; caudâ albo bifasciatâ, fasciâ terminali latâ: primâ angustâ; primariis nigrescentibus fasciâ angustâ albâ ad medium: alis spuriis maculâ albâ notatis; gutture lunulâ albâ; secundariis tectricibusque alarum maculâ fulvescente ad apicem; crisso pallidè rufescente; rostro pedibusque fuscis.

Long. tot. unc., 9¾; alæ, 6½; caudæ, 5; tarsi ¾.

Front and back of head gray, mottled with black and with little fulvous. The latter colour more abundant, and in larger markings in the interscapular region, and on the wing-coverts. The black markings give a somewhat streaked appearance to the back of head and interscapulars. On the back of throat the fulvous tint is so much pronounced, that a collar is formed which is continued under a white one round the breast. Wings : primaries brownish-black ; four external ones, with a large white mark, forming a band, at about one-third of their length from their extremities : these white marks are edged with fulvous, and the part on the outer web of the first primary, is wholly so coloured. The other primaries are marked with reddish brown, as are the secondaries and tertiaries, the marks becoming more numerous and smaller, and the colours more mottled, nearer the back. Tail: upper tail-coverts and two central feathers of tail marked like those on the back ; the black, however, forming narrow interrupted transverse bars. The pair next to these central ones have near their extremities a large white mark, but only on the inner shaft. In the three succeeding pairs, the white spot extends on both sides of the shaft, and in each pair increases somewhat in size ; so that in the external pair, the white spot is merely bordered with a very narrow, faint margin, of brown and fulvous. At about half their length, all the feathers, with the exception of the central pair, have a smaller white mark, but only on the inner side of the shaft. This mark is transverse, in the form of a band, and the white blends into fulvous on the edges of the webs. Outer web of these same external feathers, are transversely barred with black and fulvous. *Under surface.*—Chin, breast, belly, and lining of wings, dirty fulvous, with numerous narrow, irregular, transverse bars of brown. Throat with white collar, beneath which the fulvous tint is predominant, forming a kind of under collar, which is continued round the whole neck. Under tail-coverts fulvous,—tail itself appears almost black, with a great terminal white band, and a narrower one at about half its length.

Wings, an inch and a quarter shorter than the tail. Second primary, scarcely perceptibly longer than the third; the first about an eighth of an inch shorter than the second, and 8 ths longer than the fourth. Feathers on wing, with the outer webs, slightly excised.

	In.		In.
Total length	9¾	Tarsi	⁵⁄₈
Wing folded	6½	From tip of beak to rictus	1
Tail	5	Of middle toe without the claw	⸱⁷⁄₈

Habitat, Valparaiso Chile, (*August*).

This species frequents the mountains of central Chile. When bivouacking one night on the Bell of Quillota, at an elevation of 6000 feet above the sea, I heard a gentle, plaintive cry, which I was told was made by this bird. It is regarded with superstitious dread by many of the lower orders.

Mr. Gould observes, that "this species has a strong resemblance, at the first glance, with the *Caprimulgus Europæus*, but may be readily distinguished by its shorter wing, more lengthened tarsi, by a conspicuous white band across the base of the tail, and by all these feathers, except the two middle ones, having another white band near the tip." Mr. Gould then adds, as "I am quite undecided to which of the sub-genera this and the following species should belong, I leave them for the present in the restricted genus, *Caprimulgus*, although I certainly perceive in it many points of affinity to the group which inhabits the United States of North America."

2. Caprimulgus parvulus. *Gould.*

Caprimulgus parvulus, *Gould*, Proceedings of the Zoological Society, February 1837, p. 22.

C. capite intensè fusco, guttis minutis cinereis ornato; vittâ rufâ cervicem cingente; gutture scapularibusque ad marginem, secundariis ad apicem stramineis; pectore et abdomine lineis fuscis transversis; primariis nigrescentibus, tribus fasciis inæqualibus pallidè rufescentibus; caudâ fasciis pallidè fulvescentibus et fuscis ornatâ.

Long. tot. unc., 7½; *alæ*, 5; *caudæ*, 4; *tarsi*, ⅝.

Crown of head gray, with black longitudinal streaks. Back of neck with a fulvous ring, which extends round the front beneath one of white, as in the *C. bifas-*

ciatus. Back, dull gray. Interscapulars, with the central part of each feather, black, terminating in a point ; the outer part of the web being broadly fringed with a very pale fulvous, the inner with gray. Wings: primaries brown, with fulvous marks, forming three irregular transverse bars, which are scarcely visible when the wing is closed. Tail and upper tail-coverts, dull coloured, very obscurely marked with transverse bars of gray and fulvous, of different degrees of darkness. *Under surface.*—Throat white, edged with fulvous on lower side. Breast, belly, and under tail-coverts, fulvous, with numerous very narrow transverse bars of brown. The pale fulvous marks, forming interrupted bars, are more plainly seen on this than on the upper side of the tail.

Third primary, very little longer than second, and second than first. First rather longer than fourth. Extremities of wings reaching within an inch and a quarter of end of tail. End of tail more rounded than in last species.

	In.		In.
Total length	7½	Tarsi	⅝
Wings	5	Middle toe, from tip of claw to joint of foot	$\frac{9}{16}$
Tail	4	From tip of beak to rictus	1

Habitat, La Plata, (*September*).

This species is not uncommon on the wooded banks of the Parana, near Santa Fé. If disturbed, it rises from the ground, in the same inactive manner as the European species. I saw one alight on a rope diagonally, but not so completely in a longitudinal position as does the *C. Europæus*, nor transversely as other birds. Mr. Gould observes, that "this goatsucker is full a third less than the *Caprimulgus Europæus*, and is remarkable for the uniformity of its markings, having no distinct white bars, or marks, either on the wings or tail."

FAMILY.—HIRUNDINIDÆ.

1. PROGNE PURPUREA. *Boie.*

Hirundo purpurea, *Wils.*

My specimens were obtained at Monte Video, (November) and Bahia Blanca, 39° S. (September) how much further southward this species extends I do not know. Jardine says, that in North America it migrates during summer as far as the Great Bear Lake, in Lat. 66° N.; it is mentioned by M. Audubon, at New Orleans, 30° N., and by Mr. Swainson, at Pernambuco, in 8½° S.; we may, therefore, conclude that it ranges throughout both Americas, but it is not found in the Old World. Wilson describes this bird as a great favourite with the inhabitants of North America, both European and Indian, who erect boxes and other contrivances near their houses for it to build in. At Bahia Blanca, the females were beginning to lay in September, (corresponding to our March): they had excavated deep holes in a cliff of compact earth, close by the side of the larger burrows inhabited by the ground parrot of Patagonia, (*Psittacara Patagonica*.) I noticed several times a small flock of these birds, pursuing each other, in a rapid and direct course, flying low, and screaming in the manner so characteristic of the English Swift, (*Hirundo Apus*, Linn.)

2. PROGNE MODESTA. *Gould.*

PLATE V.

Hirundo concolor, *Gould*, in Proceedings of the Zoological Society.

P. nitidè cærulescenti-nigra.

Long. tot. 6 unc; alæ, $5\frac{1}{4}$; caudæ, $2\frac{3}{4}$; tarsi, $\frac{1}{2}$.

The upper and under surface has not so strongly a marked purple shade, as in the *P. purpurea*. The primaries and feathers of the tail, however, have a greenish gloss, perhaps slightly more metallic.

Tail not so deeply forked as in *P. purpurea*, which is owing to the two external feathers on each side not being so much prolonged and bent outward, as in that species. Nostrils of less size than in the latter, although the beaks differ but little. Claws and feet are much less strong, than might have been anticipated, even proportionally to the less dimensions of this species compared with the *P. purpurea*.

	Inches.		Inches.
Total length	6	Tarsi	
Wings	$5\frac{1}{4}$	Middle toe from tip of claw to joint	
Tail	$2\frac{3}{4}$		

Habitat, James Island, Galapagos Archipelago, (*October*).

Male.

This swallow was observed only on this one island of the group, and it was there very far from common. It frequented a bold cliff of lava overhanging the sea. Had not Mr. Gould characterized it as a distinct species, I should have considered it only as a small variety, produced by an uncongenial site, of the *Progne purpurea*. I can perceive no difference whatever from that bird, excepting in its less size, slenderness of limbs, and less deeply forked tail; and the latter difference may perhaps be owing to youth.

1. Hirundo Leucopygia. *Licht.*

My specimens were obtained at Port Famine, in Tierra del Fuego, (*February*), and at Valparaiso, in Chile, (*August to September*). At Port Famine they build in holes in a cliff of earth. Mr. Gould says, " were it not for the bare legs of this little Martin, I should have some difficulty in discriminating between it and the one so well known as a summer visitor in our island."

2. Hirundo frontalis. *Gould.*

H. vertice, plumis auricularibus, dorso et lunulâ pectorali nitidè cæruleo viridescentibus, notâ albâ supra nares, gulâ corporeque subtus albicantibus, crisso niveo, alis caudâque fuscis viridi tinctis, rostro nigro, pedibus intensè fuscis.

Long. tot. 4¾ unc. *alæ*, 4¾; *caudæ*, 2; *tarsi*, ½.

Upper surface, with a greenish blue metallic gloss; which can faintly be perceived on the primaries and on the tail feathers. The short feathers over each nostril white, thus forming two small white marks; those over the ridge of bill pale brown, giving together the appearance of a narrow white band over the upper mandible. Entire under surface and lining of wings pure white. Tarsi rather darker than in *H. leucopygia*.

Very slightly larger than *H. leucopygia*; upper mandible rather broader.

	Inches.		Inches.
Total length	4¾	Tail	2
Wings	4¾	Tarsi	½

Habitat, Monte Video, (*November*).

Mr. Gould says, "this species is closely allied both to the common martin, and to the last species; from the former bird, however, its bare legs at once distinguish it, while it differs from the latter in being rather larger in size, in having an obscure white mark on the forehead, at the base of the bill, and in having the metallic lustre of the upper surface deep steel green, instead of purple, which is the prevailing colour of both *Hirundo leucopygia* and *H. urbica*."

It is abundant on the northern bank of the Plata, and more common than the *H. purpurea*, which frequents the same localities. It probably replaces on the eastern side of the continent, the *H. leucopygia* of Chile.

3. Hirundo cyanoleuca. *Vieill.*

It is nearly allied to the two latter species, but is readily distinguished from them by the absence of the white rump. I procured specimens in September, both from Valparaiso, and from Bahia Blanca (North Patagonia). At the latter place it built in holes in the same bank of earth with *P. purpurea*.

Cypselus unicolor. *Jard.*

C. unicolor. *Jard. et Selby,* Illust. Ornith. pl. 83.

I obtained a specimen of this bird from St. Jago, Cape de Verd Islands. (September).

It more resembled a swallow than a swift in the manner of its flight. I only saw a few of them. Insects occur so scantily over the bare and parched plains of basaltic lava, which compose the lower parts of the island of St. Jago, that it is surprising how these birds are able to find the means of subsistence.

Family.—HALCYONIDÆ.

Halcyon erythrorhyncha, *Gould,* Proc. Zool. Soc. 1837.

Alcedo Senegalensis var. β, *Lath.*

In January, during the first visit of the Beagle to St. Jago, in the Cape de Verd Islands, these birds were numerous. But in our homeward voyage, in the beginning of September, I did not see a single individual. As Mr. Gould informs me it is an African species; it is probably only a winter visitant to this archipelago. It lives in numbers in the arid valleys in the neighbourhood of Porto Praya, where it may be generally seen perched on the branch of the castor oil plant. I opened the stomachs of several, and found them filled with the wing cases of Orthopterous insects, the constant inhabitants of all sterile countries; and in the craw of one there was part of a lizard. It is tame and solitary; its flight is not swift and direct like that of the European kingfisher. In these respects, and especially in its abundance in dry rocky valleys where there is not a drop of water, it differs widely from the habits of the allied genus Alcedo; although certainly it abounded more in those valleys where streamlets occurred. This Halcyon was the only brilliantly coloured bird which I saw on the island of St. Jago.

1. CERYLE AMERICANA, *Boie.*

Alcedo Americana, *Gmel.*

This Kingfisher is common on the banks of the Parana. It frequents the borders of lakes and rivers, and sitting on the branch of a tree, or on a stone, it thence takes short flights, and dashes into the water to secure its prey. Its manner of flying is neither direct nor rapid, which character is so remarkable in the flight of the European species; but it is weak and undulatory, and resembles that of the soft-billed birds. It often arrests itself suddenly in its course, and hovers over the surface of the water, preparatory to darting on some small fish. When seated on a twig it constantly elevates and depresses its tail; and as might have been expected from its figure, it does not sit in the stiff upright position so peculiar to the European Kingfisher. Its note is not unfrequently uttered: it is low, and like the clicking together of two small stones. I was informed that it builds in trees. The internal coating of the stomach is of a fine orange colour. Mr. Gould has seen specimens of this bird from Mexico; it enjoys, therefore, a very wide range.

2. CERYLE TORQUATA, *Bonap.*

Alcedo torquata. *Gmel.*
Ispida torquata. *Swain.*

This bird is common in the south part of Chile, in Chiloe, the Chonos Archipelago, and on the whole west coast, as far as the extreme southern parts of Tierra del Fuego. In these countries, it almost exclusively frequents the retired bays and channels of the sea with which the land is intersected; and lives on marine productions. I opened the stomach of one, and found it full of the remains of crustaceæ, and a part of a small fish. It occurs likewise in La Plata, and is very common in Brazil, where it haunts fresh water. It is said *(Dict. Class. d'Hist. Nat.)* to occur in the West Indian islands; it has, therefore, a wider range (from the equatorial region to the neighbourhood of Cape Horn) than the *Ceryle Americana.*

FAMILY.—MUSCICAPIDÆ. *Vieill.*

SUB-FAM.—TYRANNINÆ. Sw.

SAUROPHAGUS SULPHURATUS. *Swains.*

Lanius sulphuratus. *Gmel.*
Tyrannus magnanimus. *Vieill.* Ency. Meth. p. 850.
Tyrannus sulphuratus. *D'Orb. et Lafr.* Mag. de Zool. 1837, p. 42.

The habits of this bird are singular. It is very common in the open country, on the northern banks of the Plata, where it does not appear to be a bird of passage. It obtains its food in many different methods. I have frequently observed it, hunting a field, hovering over one spot like a hawk, and then proceeding on to another. When seen from a short distance, thus suspended in the air, it might very readily be mistaken for one of the rapacious order; its stoop, however, is very inferior in force and rapidity. At other times the Saurophagus haunts the neighbourhood of water, and there, remaining stationary, like a kingfisher, it catches any small fish which come near the margin. These birds not unfrequently are kept, with their wings cut, either in cages or in court-yards. They soon become tame, and are very amusing from their cunning odd manners, which were described to me, as being similar to those of the common magpie. Their flight is undulatory, for the weight of the head and bill appears too great for the body. In the evening the Saurophagus takes its stand on a bush, often by the road-side, and continually repeats, without change, a shrill and rather agreeable cry, which somewhat resembles articulate words. The Spaniards say it is like the words, "Bien te veo" (I see you well), and accordingly have given it this name.

MUSCIVORA TYRANNUS. *G. R. Gray.*

Muscicapa Tyrannus. *Sw.*
Tyrannus Savana. *Vieill. Bonap.* Am. Orn. pl. 1. f. 1.

This species belongs to Mr. Swainson's genus Milvulus (more properly Milvilus,) but which name Mr. G. R. Gray has altered to Muscivora as the latter was proposed for *Musc. forficata* as far back as 1801, by Lacepède.

It is very common near Buenos Ayres; but I do not recollect having seen many in Banda Oriental. It sits on the bough of a tree, and very frequently on the ombu, which is planted in front of many of the farm houses, and thence takes short flights in pursuit of insects. From the remarkable structure of its tail, the inhabitants of the country call it scissor-tail; a name very well applied from the manner in which it opens and shuts the forked feathers of its tail. Like all birds thus constructed, (of which the frigate bird offers a most striking example), it has the power of turning very shortly in its flight, at which instant it opens and shuts its tail, sometimes, as it appears, in a horizontal and sometimes in a vertical plane. When on the wing it presents in its general appearance a caricature likeness of the common house swallow (*Hirundo rustica*). The Muscivora, although unquestionably belonging to the family of Muscicapidæ manifests in its habits an evident relationship with birds of the fissirostral structure.

Sub-Gen. PYROCEPHALUS, Gould.

Muscicapa.   *Auct.*
Muscipeta.   *Cuv.*
Tyrannula.   *Swain.*

Rostrum capite brevius, rectum, depressum, basi setis numerosis nigris obsessum; mandibulâ superiore emarginatâ, inferiorem obtegente; naribus rotundatis patulis. Caput subcristatum. Alæ longæ; remige prima secundum tertiamque longissimas subæquales fere æquante. Tarsi mediocres, anticè scutellati; digitis lateralibus inæqualibus, exteriore longiore. Cauda mediocris quadrata.

Mr. Gould observes, that "the males of nearly all the members of this group (which may be considered either as a distinct genus or sub-genus of Myiobius), have the crown of the head and greater part of the under surface scarlet. Four species were obtained.—*Pyrocephalus parvirostris, (Gould)*, and *Muscicapa coronata, (Auct.)*, may be taken as types.

1. Pyrocephalus parvirostris. *Gould.*

Plate VI.

Le Churrinche, *Azara*. No. 177.

P. suprà fuscus; capite et subtus nitidè puniceis; rectricibus exterioribus tectricumque et secundariorum apicibus griseo-marginatis.

Long. tot. $5\frac{1}{12}$ unc.; *alæ*, $13\frac{1}{12}$; *caudæ*, $2\frac{1}{12}$; *tarsi*, $\frac{7}{12}$; *rost.* $\frac{6}{12}$.

Crown of the head, crest, and all the under surface, bright scarlet; the remainder of the plumage, deep brown; the outer tail-feathers on each side, and the edges of the secondaries and wing-coverts, margined with grey.

Habitat, La Plata, (*October.*)

This species differs from *Pyr. coronatus* or *Musicapa coronata*, of authors, chiefly in its size; in other respects it is very similar. The admeasurements of the latter, for comparison (as given me by Mr. G. R. Gray), are: total length, 5 inches and 8 lines; bill, between 9 and 10 lines; wings, 3 inches and 2 lines; tail, 2 inches and 7 lines; tarsi, 7 or 8 lines.

During the summer, this bird was common both near Buenos Ayres and Maldonado; but at the latter place, I did not see one in the months of May, June, July, (winter) and therefore, no doubt it is a bird of passage, migrating southward during the summer from Brazil. The birds of this and the allied genera, correspond very closely in their habits to certain of the Sylviadæ of Europe; some of the species frequenting bushes, like the black-cap, (*Sylvia atricapilla*); others more usually the ground, as the robin (*Sylvia rubecula*) or hedge-sparrow (*Accentor modularis*). Another group (*Synallaxis, &c.*) represent those European Sylviæ, which frequent reeds.

2. Pyrocephalus obscurus. *Gould.*

P. lividus rufotinctus; præcipuè in fronte ventreque.

Long. tot. $5\frac{2}{12}$ unc.; *alæ*, $3\frac{2}{12}$; *caudæ*, $2\frac{1}{12}$; *tarsi*, $\frac{7}{12}$; *rost*. $\frac{8}{12}$.

All the plumage chocolate-brown, tinged with red, the latter colour predominating on the forehead and lower part of the abdomen; bill and tarsi, black.

A single specimen was obtained, and it would appear to be either an immature bird or a female.

Habitat, Lima, Peru. (*August.*)

3. Pyrocephalus nanus. *Gould.*

Plate VII.

P. fuscus; rectricum exteriorum marginibus omniumque et secundariorum apicibus nitidè griseo-brunneis.

Femina, brunnea; gutture griseo-albo; corpore subtus pallidè flavescente; pectoris laterumque plumis in medio brunneo-striatis.

Long. tot. $4\frac{1}{2}$ unc.; *alæ*, $2\frac{9}{12}$; *caudæ*, $2\frac{0}{12}$; *tarsi*, $\frac{8}{12}$; *rostri*, $\frac{6}{12}$.

Male.

Crown of the head, crest, and all the under surface, scarlet; back, wings, and tail, sooty-brown; the external margin of the outer tail feathers, and the tips of all, light greyish brown; bill and tarsi, black.

Female.

All the upper surface, wings, and tail, brown; throat, greyish white; the remainder of under surface, pale buff, the feathers of the chest and flanks, with an obscure fine stripe of light brown down the centre.

Habitat, Galapagos Archipelago. (*September*.)

There is nothing remarkable in the habits of this bird. It frequents both the arid and rocky districts near the coast, and the damp woods in the higher parts of several of the islands in the Galapagos Archipelago.

4. PYROCEPHALUS DUBIUS. *Gould*.

P. minor, lividus; fronte, superciliis corporeque subtus stramineis; tectricibus stramineo marginatis.

Long. tot. $4\frac{1}{12}$ unc; *alæ*, $2\frac{4}{12}$ *caudæ*, $1\frac{9}{12}$; *tarsi*, $\frac{7}{12}$; *rost*.

Forehead, stripe over the eye, and all the under surface pale buff; back of the neck and upper surface chocolate brown; greater and lesser wing coverts margined with buff.

Habitat, Galapagos Archipelago, (*September*).

From the appearance of this bird when alive, although closely resembling *P. nanus*, I entertained no doubt that it was a distinct species. Mr. G. R. Gray informs me that there is a specimen of a male in the British Museum, which differs from the male of the precedent species, in having the upper colour of a decided brown, and the external margins of the outer tail feathers and tips of the secondaries rather reddish white; also in size as stated by Mr. Gould.

MYIOBIUS. *G. R. Gray*.

TYRANNULA. *Swains*.

Mr. Gould had adopted for the following species Mr. Swainson's generic appellation of *Tyrannula*, but Mr. G. R. Gray has pointed out, that as Tyrannulus was proposed and published eleven years before, namely in 1816, by Vieillot, it becomes necessary to change the former name, and therefore he proposes *Myiobius*.

1. Myiobius albiceps. *G. R. Gray.*

Muscipeta albiceps. *D'Orb. et Lafr.* Mag. de Zool. 1837, p. 47.

This bird is not uncommon in Tierra del Fuego, and along the western coast of the southern part of the continent, where the land is covered with trees; it is occasionally found near Valparaiso in central Chile ; and likewise in Banda Oriental on the banks of the Plata, where the country is open, from all of which places I procured specimens. At Port Famine and in the islands of the Chonos Archipelago, it inhabits the gloomiest recesses of the great forests. It generally remains quietly seated high up amongst the tallest trees, whence it constantly repeats a very plaintive, gentle whistle, in an uniform tone. The sound can be heard at some distance, yet it is difficult to perceive from which quarter it proceeds, and from how far off; and I remained in consequence, for some time in doubt, from what bird it proceeded.

2. Myiobius auriceps.

Tyrannula auriceps. *Gould*, MS.

M. rufus; capite cristato nitidè flavo; plumarum apicibus brunneis; alis brunneis, secundariarum marginibus tectricumque apicibus rufis; caudâ pallidè brunneâ, plumarum externarum marginibus externis pallidioribus; gutture corporeque subtus pallidè flavescenti-albis; plumis singulis fasciâ centrali brunneâ.

Long. tot. $5\frac{3}{12}$ unc; *alæ*, $2\frac{5}{12}$ *caudæ*, $2\frac{6}{12}$ *tarsi*, $\frac{9}{12}$ *rost.* $\frac{7}{12}$.

All the upper surface rufous; the basal portion of the coronal feathers yellow ; tail uniform light brown, the external margin of the outer feathers lighter ; wings brown, the external margin of the secondaries and the tips of the greater and lesser wing-coverts rufous ; throat and all the under surface pale buffy white, each feather having a brown mark down the centre ; bill brown ; feet black.

Habitat, Buenos Ayres, La Plata, (*August*).

This bird is about the size of a sparrow. It is nearly allied to *Tyrannula ferruginea* of Swainson and *M. cinnamonea* of D'Orbig. and Lafr.

3. Myiobius parvirostris.

Tyrannula parvirostris, *Gould*, MS.

M. suprà rufobrunneus; pileo, nuchâ humerisque obscurè olivaceo-brunneis; alis brunneis, primariarum et secundariarum marginibus exterius angustè tectricumque latè ferrugineis; caudâ guttureque griseo-brunneis; pectore abdomineque flavescenti brunneis.

Long. tot. 4 1/8 unc.; alæ, 2 6/12; caudæ, 2 9/12; tarsi, 9/12; rost. 6/12.

Crown of the head, back of the neck, and shoulders, dark olive brown; back and upper tail coverts rufous brown; wings brown; the external edges of the primaries and secondaries finely, and the greater and lesser wing coverts broadly margined with ferruginous; tail uniform greyish brown; throat brownish grey; chest and abdomen sandy brown; upper mandible dark brown; under mandible yellowish brown; feet blackish brown.

Habitat, Tierra del Fuego, Chile, and La Plata.

This bird inhabits the forests of Tierra del Fuego, and as I procured specimens of it in the beginning of winter (June), it probably remains throughout the year in the extreme southern part of South America. Other specimens were procured on the banks of the Plata, and near Valparaiso in Chile; it has therefore a wide range.

4. Myiobius magnirostris.

Plate VIII.

Tyrannula magnirostris. *Gould*, MS.

M. Fæm. Suprà olivaceo-brunnea; caudâ brunneâ; rectricum externarum marginibus griseo-brunneis; gutture pectoreque olivacco griseis; abdomine caudæque tectricibus inferioribus pallidè flavis; alis saturatè brunneis, secundariis tectricibusque latè griseo marginatis.

Long tot. 5 1/12; alæ, 2 8/12; caudæ, 2 6/12; tarsi, 11/12; rost. 8/12.

Crown of the head and back olive brown; tail brown; the external margins of the two outer feathers greyish brown; throat and chest olive grey; abdomen and under tail coverts very pale citron yellow; wings dark brown; secondaries, greater and lesser wing coverts broadly margined with grey; bill and feet black.

Habitat, Chatham Island, Galapagos Archipelago (*October*).

This bird and the *Pyrocephalus nanus*, inhabit the same island. Not very uncommon.

Genus.—SERPOPHAGA. *Gould.*

Rostrum capite multò brevius, rectum, subdepressum; tomiis rectis; mandibulâ superiore subemarginatâ; naribus basalibus, lateralibus, pilis mollibus anticè versis partim tectis. Alæ breves, concavæ, remige quartâ longissimâ. Cauda longiuscula subrotundata. Tarsi mediocres squamis duris annulati; digitis parvis, postico mediano breviore, lateralibus æqualibus, exteriore cum mediano usque ad articulum priorem connatum.

1. SERPOPHAGA PARULUS. *Gould.*

Muscicapa parulus, *Kittlitz*, Mem. L'Acad. Imp. des Sci. St. Peters. 1831. 1. p. 190. Pl. 9.
Sylvia Bloxami, *Gray's* Zool. Misc. 1831. p. 11.
Culicivora parulus, *D'Orbig.* & *Lafr.* Mag. de Zool. 1837, p. 57.

This bird is common in central Chile, in Patagonia, and although found in Tierra del Fuego, it is not numerous there. Its specific name is very well chosen, as I saw no bird in South America whose habits approach so near to those of our tom-tits (*Parus*). It frequents bushes in dry places, actively hopping about them, and sometimes repeating a shrill cry; it often moves in small bodies of three and four together. In August I found the nest of one in a valley in the Cordillera of central Chile; it was placed in a bush and was simply constructed.

2. SERPOPHAGA ALBO-CORONATA. *Gould.*

S. supra olivaceo-brunnea, subtus pallidè flava; pileo nigrescenti brunneo, in hoc plumarum basibus lineâque supra oculos albis; alis nigrescenti brunneis, primariis angustè olivaceo marginatis, tectricibus latè olivaceo-griseo marginatis, gutture griseo.

Long. tot. $4\frac{3}{12}$; alæ, 2; caudæ, 2; tarsi, $\frac{7}{12}$; rost. $\frac{4}{12}$.

A stripe of white from the nostrils over each eye; crown of the head brown, the base of all the feathers pure white; back of the neck, back and upper tail coverts olive brown; wings blackish brown, the external edges of the primaries finely margined with olive, and the greater and lesser wing coverts largely tipped with olive grey; tail uniform brown; throat grey; abdomen and under tail coverts pale citron yellow; bill and feet brown.

Habitat, Maldonado, La Plata, (*June*).

This bird, like the last species, generally moves in very small flocks. Its habits, I presume, are also very similar; for I state in my notes that it closely approaches to our tit-mice in general manners and appearance.

3. SERPOPHAGA NIGRICANS. *Gould.*

Sylvia nigricans, *Vieill.*
Tachuris nigricans, *D'Orbig. & Lafr.* Mag. de Zool. 1837. p. 55.
Le Petit Tachuris noirâtre, *Azara,* No. 167.

This bird is common in the neighbourhood of Maldonado, on the banks of the Plata. It generally frequents the borders of lakes, ditches, and other moist places; but is related in its general manners with the last species. It often alights on aquatic plants, growing in the water. When seated on a twig it occasionally expands its tail like a fan.

SUB.-FAM.—TITYRANÆ. (PSARIANÆ, *Sw.*)

PACHYRAMPHUS, *G. R. Gray.*

Pachyrhynchus, *Spix.*

1. PACHYRAMPHUS ALBESCENS.

Pachyrhynchus albescens, *Gould, MS.*

PLATE XIV.

P. olivaceo-griseus; alis nigrescenti brunneis, albescenti marginatis; gutture corporeque subtus griseo-albis; alarum tectricibus inferioribus pallidè sulphureis.

Long. tot. $5\frac{3}{12}$ unc.; *alæ*, $2\frac{7}{12}$; *caudæ*, $2\frac{6}{12}$; *tarsi*, $\frac{8}{12}$; *rost.* $\frac{7}{12}$.

Head and all the upper surface olive grey; wings blackish brown, the coverts and secondaries broadly margined with dull white; primaries narrowly margined with greyish white; tail blackish brown, the external web of the outer feather white; under surface of the shoulder pale sulphur yellow; throat and under surface greyish white; bill and feet black.

Habitat, Buenos Ayres.

The generic name of Pachyrhynchus *Spix,* is changed by Mr. G. R. Gray, to *Pachyramphus,* as the former word is used in entomology.

2. Pachyramphus minimus.

Pachyrhynchus minimus, *Gould, MS.*

Plate XV.

P. rufo brunneus; capite guttureque brunneo-nigris; plumarum basibus albis; alis caudáque brunneis, plumis flavescenti-albo marginatis; colli lateribus, fasciâ pectorali hypochondriisque fulvis; jugulo ventreque pallidè flavescentibus.

Long. tot. $3\frac{7}{12}$; alæ, $1\frac{10}{12}$; cauda, $1\frac{7}{12}$; tarsi, $\frac{8}{12}$; rost. $\frac{6}{12}$.

Crown of the head, sides of the face and throat blackish brown, each feather white at the base; back of the neck black, and upper tail coverts rufous brown; wings and tail dark brown, each feather margined with sandy white; sides of the neck, under surface of the shoulder, band across the chest and flanks reddish fawn colour; lower part of the throat, and centre of the abdomen very pale buff; bill and feet blackish brown.

Habitat, Monte Video, (*November*).

Sub-Fam.—FLUVICOLINÆ, Swain.

Alecturus guirayetupa. *Vieill.* Dict.

Muscicapa psalura, *Temm.*, Pl. Col. t. 286 and 296.
——— risoria, *Vieill.*, Gal. des Ois. Pl. 131.
Yetapa psalura, *Less.*, Tr. d'Orn. i. p. 387.
Le Guirayetupa, *Azara*, No. 220.

This bird is not uncommon on the open grassy country near Maldonado on the banks of the Plata. It sits generally on the top of a thistle; from which it makes short flights and catches its prey in the air. The two long feathers in its tail appear quite useless to it. It sometimes feeds on the ground. In the stomach of one which I opened there was a spider (*Lycosa*), and some Coleoptera.

1. Lichenops perspicillatus. *G. R. Gray.*

Sylvia perspicillata, *Gmel.*
Œnanthe perspicillata, *Vieill.*
Ada Commersoni, *Less.*
Perspicilla leucoptera, *Swains.*, Nat. Libr. x. Flyc. p. 105, Pl. 9.
Fluviola perspicillata, *D'Orb. & Lafr.*, Mag. de Zool. 1837, p. 59.
Le Clignot ou Lichenops, *Comm., Sunder.*
Le Bec d'argent, *Azara*, No. 228.

This bird belongs to the sub-genus, *Perspicilla*, of Mr. Swainson; but as Mr.

G. R. Gray has pointed out that Commerson had previously considered it the type of his genus, *Lichenops*, we have been induced to prefer the latter as the oldest name. It is common in the neighbourhood of the Plata, and across the Pampas, as far as Mendoza on the eastern foot of the Andes; it has not, however, crossed those mountains and entered Chile. It usually sits on the top of a thistle, and like our common fly-catchers (*Muscicapa grisola*), takes short flights in pursuit of insects; but does not, like that bird, return to the same twig. It feeds, also, occasionally on the turf: in the stomach of some which I opened, I found Coleopterous insects, chiefly Curculionidæ. Beak, eye-lid, and iris, beautiful primrose yellow.

2. Lichenops erythropterus. *Gould.*

Plate IX.

L. suprà nigrescenti-brunneus, plumis rufo-marginatis; primariis secundariisque castaneis, apicibus pogoniæque externæ dimidio apicali brunneis; gutture corporeque subtus cervinis; pectore brunneo-marginato.

Long. tot. 6 unc.; *alæ*, 3; *caudæ*, $2\frac{3}{8}$; *tarsi*, 1; *rostri*, $\frac{9}{17}$.

All the upper surface and tail blackish brown, each feather margined with rufous; primaries and secondaries reddish chesnut, their tips and their external webs for half their length from the tip, brown; tertiaries, greater and lesser wing-coverts dark-brown, each feather margined with reddish buff; throat, and all the under surface, fawn colour; the chest spotted with brown; base of the bill, and chiefly of the lower mandible, as well as the iris, bright yellow; eye-lid, blackish yellow; feet, dark brown.

Habitat, Banks of the Plata.

This bird is not very common. It frequents damp ground, where rushes grow, on the borders of lakes. It feeds on the ground and *walks*. It is certainly allied in many respects with the foregoing species, but in its power of walking, and in feeding on the ground, there is a marked difference in habits. As it has lately been described (Swainson's Nat. Libr. Ornith. x. p. 106.) as the female of the *L. perspicillatus*, I will here point out some of its chief distinguishing characters. Its beak is slightly more depressed, but with the ridge rather more plainly pronounced. In the *L. perspicillatus*, the upper mandible is entirely yellow, excepting the apex; in the *L. erythropterus*, it is entirely pale brown, excepting the base. The eyelid in the former is bright primrose yellow, in the latter blackish yellow. The tail of *L. erythropterus* is squarer and contains only ten feathers instead of twelve: the wing is $\frac{7}{10}$ of an inch shorter, and the secondaries relatively to the primaries are also shorter. The red colour on the primaries represents, but does not correspond with, the white on the black feathers of *L. perspicillatus;* and the secondaries in the two birds are quite differently marked. In *L. erythropterus*, the third, fourth, and fifth primaries are the longest, and are equal to each other; the second is only a little shorter than the third. In *L. perspicillatus* the third is rather shorter than the fourth and fifth; and the second is proportionally shorter relatively to the third, so that the outer part of the wing in this species is more pointed than in *L. erythropterus*. The hinder claw in the latter is only in an extremely small degree straighter than in the former; and this, considering that the *L. perspicillatus* is generally perched, and when on the ground, can only hop; and that the *L. erythropterus* feeds there entirely, and walks, is very remarkable.

1. FLUVICOLA ICTEROPHRYS.   *D'Orb. & Lafr.* Mag. de Zool. 1837. p. 59.

Muscicapa icterophrys, *Vieill.* Encyc. Meth. p. 832.
Le Suiriri noirâtre et jaune, *Azara*, No. 183.

Specimens were found by me both at Monte Video and at Maldonado, on the banks of the Plata. I found Coleoptera in their stomachs.

2. FLUVICOLA IRUPERO.   *G. R. Gray.*

Tyrannus Irupero, *Vieill*, Ency. Meth. p. 856.
Muscicapa mœsta, *Licht.* Cat. p. 54.
Muscicapa nivea, *Spix*, Av. pl. 29. f. 1.
Pepoaza nivea, *D'Orb. & Lafr.* Mag. de Zool. 1837. p. 62.
Irupero, *Azara*, No. 204.

This elegant bird, which is conspicuous amongst most land species by the whiteness of its plumage, is found, though not commonly, (in November) in Banda Oriental; whilst near Santa Fé, three degrees of latitude northward, it was common during the same time of year. It is rather shy, generally perches on the branches of bushes and low trees.

3. FLUVICOLA AZARÆ.   *Gould.*

PLATE X.

F. alba; alis, caudâ caudæque tectricibus atris, his albo-marginatis; primariis flavescenti-albis, basibus apicibusque nigris; rostro pedibusque atris.

Long. tot. $8\frac{3}{12}$ unc.; alæ, $4\frac{9}{12}$; caudæ, $4\frac{3}{12}$; tarsi, 1; rost. 1.

Head, all the upper and under surface white; wings and tail black; tail coverts black margined with white; primaries broad and crossed near their extremity with sulphur white, and tipped with brown; bill and legs black.

Habitat, banks of the Plata.

This bird is very common in the neighbourhood of Maldonado, where it frequents the open grassy plains. It sits on the top of a thistle, or on a twig, and catches the greater part of its food on the wing. It is generally quiet in its movements and silent. Mr. Gould remarks, that he finds "nearly all the species of this peculiar group to differ remarkably in the structure of their wings and tail, while in all other respects they closely resemble each other both in form and habit; I have, therefore, hesitated to separate them into so many genera. I have assigned the present species to Mr. Swainson's subgenus *Fluvicola*, considering that differences in the form of one organ alone would not be sufficient grounds for the institution of a new genus among such closely allied species; the present bird evidently leads off to *Tænioptera*, a genus proposed many years since, by the Prince of Musignano for the *Pepoazas* of Azara.

"This species is closely allied to, if not identical with the *Pepoaza Dominicana* of Azara, but as there is a degree of obscurity in his description, which causes some doubt on this point, I have considered it better to pay a just tribute of respect to that zealous labourer in the field of natural science, by assigning his name to this very elegant bird."

1. Xolmis coronata. *G. R. Gray.*

Tyrannus coronatus, *Vieill.* Ency. Meth. p. 885.
Muscicapa vittiger, *Licht.* Cat. p. 54.

My specimen was obtained on the wooded banks of the Parana, near Santa Fé, in Lat. 31° S.

Boie's name of Xolmis is adopted by Mr. G. R. Gray, as it was proposed some five years anteriorly to that of the Prince of Musignano's.

2. Xolmis nengeta. *G. R. Gray.*

Lanius nengeta, *Linné*, 1. p. 135. 7.
Tyrannus nengeta, *Swains.* Journ. Sci. xx. p. 279.
Fluvicola nengeta, *Swains.* Nat. Libr. Fly-catchers, p. 102. pl. 8.
Tyrannus pepoaza, *Vieill.* Ency. Meth. p. 855.
Muscicapa polyglotta, *Licht. Spix.* II. pl. 24.
Tyrannus polyglottus, *Cuv.*
Le Pepoaza proprement dit, *Azara*, No. 201.

My specimen was procured at Maldonado, north bank of La Plata, where it is not common. Its habits in many respects are like those of the *Fluvicola Azaræ*; it appears to catch its prey on the wing. Iris bright red.

3. Xolmis variegata. *G. R. Gray.*
Plate XI.

Pepoaza variegata, *D'Orb. & Lafr.* Mag. de Zool. 1837. p. 63. Voy. dans l'Amér. Mèr. Orn. pl. 39. f. 2.
Tænioptera variegata. On plate.

This bird feeds in small flocks, often mingled with the icteri, plovers, and other birds on the ground. Its manner of flight and general appearance never failed to call to my recollection our common fieldfares (*Turdus pilaris*, Linn.) and I may observe that its plumage (in accordance with these habits) is different from that of the rest of the genus. I opened the stomachs of some specimens killed at Maldonado, and found in them seeds and ants. At Bahia Blanca I saw these birds catching on the wing large stercovorous Coleoptera; in this respect it follows the habits, although in most others it differs from those of the rest of its tribe. Iris rich brown.

4. Xolmis pyrope. *G. R. Gray.*

Muscicapa pyrope, *Kitlitz.* Mem. l'Acad. Imp. des Sci. St. Peters. 1831. p. 191. pl. 10. Vögel von Chili, pl. 10. p. 19.
Pepoaza pyrope, *D'Orb. & Lafr.* Mag. de Zool. 1837. p. 63.

This bird is not uncommon near Port Famine in Tierra del Fuego, and along the whole western coast (at Chiloe specimens were obtained) even as far north as the desert valley of Copiapó. In the thickly wooded countries of Tierra del Fuego and Chiloe, where it is more common than further northward, it generally takes its station on the branch of a tree, on the outskirts of the forest. When thus perched, usually at some height above the ground, it sharply looks out for insects passing by, which it takes on the wing. Iris scarlet. It builds a coarse nest in bushes. Egg perfectly white, pointed oval; length one inch, breadth ·76 of an inch.

Genus.—AGRIORNIS. *Gould.*

Tyrannus, *Eyd. & Gerv.*
Pepoaza, *D'Orb. & Lafr.*

Rostrum longitudine capitis, rectum, forte, compressum, abruptè deflexum, emarginatum; tomiis rectis integris; naribus basalibus, lateralibus, rotundis, patulis; rictu pilis rigidiusculis obsesso. Alæ mediocres, remige primâ longâ, tertiâ quartâque æqualibus, longissimis. Cauda mediocris, quadrata. Tarsi longi, fortes, squamis crassis annulati; digito ungueque postico mediano breviore, lateralibus æqualibus, liberis.

Mr. Gould observes that the members of this genus are remarkable for their robust form and for their strength and magnitude of their bills; and their habits strictly accord with their structure, as they are fierce and courageous.

The species are closely allied to those of the preceding genus.*

1. Agriornis gutturalis. *Gould.*

Tyrannus gutturalis, *Eyd. & Gerv.* Voyage de la Fav. Ois. dans Mag. de Zool. 1836. pl. 11.

Pepoaza gutturalis, *D'Orb. & Lafr.* Mag. de Zool. 1837. p. 64.

My specimens were obtained near Valparaiso in Chile. I saw it as far north as the valley of Copiapó. I was assured by the inhabitants that it is a very fierce bird, and that it will attack and kill the young of other birds.

2. Agriornis striatus. *Gould.*

A. Fœm. intensè olivaceo-brunnea; alis caudâque fuscis, utriusque plumis marginibus apiceque pallidè brunneis; rectricum externarum pogoniâ externâ albâ; gutture facieque lateribus albis, his nigrostriatis; pectore hypochondriisque olivaceo-brunneis; ventre crissoque flavescentibus.

Long. tot. 10 unc.; *alæ*, $4\frac{9}{12}$; *caudæ*, $4\frac{3}{12}$; *tarsi*, $1\frac{3}{12}$; *rostri*, $1\frac{9}{12}$.

Head, and all the upper surface dark olive brown; wings and tail dark brown, each feather margined and tipped with pale brown, and the outer web of the external tail-feather, white; throat, and sides of the face, white, striated with

---

* Perhaps to this genus belong *Muscicapa thamnophiloides* and *cinerea*, figured by Spix, in his Aves, pl. 26. f. 1 and 2. *G. R. Gray.*

black; breast and flanks olive brown; centre of the abdomen and under tail-coverts, buff; bill, horn colour; feet, black.

Habitat, Santa Cruz, Patagonia. (*April.*)

I am not aware of any difference in habits between this species, and the following (*A. micropterus*); and the country inhabited by it is similar. From these circumstances I am induced to suspect, that it is the same species in an immature state.

3. AGRIORNIS MICROPTERUS. *Gould.*

Plate XII.

M. pallidè brunneus, subtus flavescenti-albus; alarum caudæque plumis griseo-marginatis; gutturis albis, brunneo-marginatis.

Long. tot. $9\frac{3}{12}$ unc.; *alœ*, $4\frac{1}{2}$; *caudæ*, $2\frac{7}{8}$; *tarsi*, $1\frac{3}{12}$; *rostri*, $1\frac{3}{8}$.

Head, all the upper surface, wings and tail, pale brown, each feather of the wings and tail margined with greyish brown; throat, white, striated with dark brown; the remainder of the under surface, buffy white; bill, dark horn colour; feet brown.

Habitat, Port Desire, and St. Julian, Patagonia. (*January*).

These birds frequent the wild valleys in which a few thickets grow. They generally take their stand on the upper twigs. They are shy, solitary, and not numerous. Mr. G. R. Gray considers the two specimens which were obtained to be immature, and that one is a full-fledged young, and the other a nestling of the *Agr. striatus.*

4. AGRIORNIS MARITIMUS. *G. R. Gray.*

PLATE XIII.

Pepoaza maritima, *D'Orb. et Lefr.*, Mag. de Zool. 1837, p. 65.
Agriornis leucurus. *Gould's MSS.*, and on Pl. xiii.

Inhabits the coast of Patagonia. It is a scarce, shy, solitary bird, frequenting the valleys in which thickets grow, but often feeding on the ground. In the interior plains of Patagonia, on the banks of the Santa Cruz, I several times saw it chasing beetles on the wing, in a peculiar manner, half hopping and half flying; when thus employed, it spreads its tail, and the white feathers in it are displayed in a very conspicuous manner. I also met with this species in the lofty and arid valleys on the eastern side of the Cordillera of Central Chile, and likewise at Copiapó.

Family.—LANIADÆ.

Sub-Fam.—LANIANÆ, Swains.

Cyclarhis Guianensis, *Swains.*

C. Guianensis, *Swains.*, Ornith. Draw. Pl. 58. ♀
Tanagra Guianensis, *Gmel.*
Laniagra Guyanensis, *D'Orb. et Lafr.*
Falcunculus Guianensis, *Swains.*, (1837.)
Le Sourciroux, *Levaill.* Ois. D'Afr. Pl. 76. f. 2.

My specimen was obtained at Maldonado, in the latter end of May. I did not see another during my residence there. In its stomach were Coleoptera.

Sub-Fam.—THAMNOPHILINÆ.

Thamnophilus doliatus, *Vieill.*

Lanius doliatus, *Linné.*

My specimen was obtained at Maldonado, where it is not very common. It generally frequents hedge-rows. Cry rather loud, but plaintive and agreeable. Iris, reddish orange; bill, blue, especially base of lower mandible. I observed individuals (females?) in which the black and white bands on the breast were scarcely visible, and even those on the under tail-coverts but obscurely marked.

Family.—TURDIDÆ.

1. Turdus rufiventer. *Licht.*

T. rufiventer, *Licht.* Cat. p. 38.
———— *Vieill.* Ency. Meth. p. 639 ?
———— *Spix,* Av. Sp. Nov. tom. 1. p. 70. t. lxviii.
———— *D'Orb. et Lafr.* Voy. de l'Amer. Mer. Av. p. 203.
Grive rousse et noirâtre, *Azara,* No. 79.
Turdus Chochi, *Vieill.* Ency. Meth. p. 639.
———— *D'Orb. et Lafr.* Mag. de Zool. 1835. p. 17.
T. leucomelas, *Vieill.* Ency. Meth. 644.
T. albiventer, *Spix,* Av. Sp. Nov. tom. 1. p. 70. t. lxix. f. 1. m. 2 fem.
La grive blanche et noirâtre, *Azara,* No. 80.

The white-bellied thrush, described under the three latter synonyms, according to M. D'Orbigny, (p. 203 of the ornithological part of his work), is the female of the *T. rufiventer*. My specimens were obtained at Maldonado and the Rio Negro, which latter place, in 41°, is its most southern limit: Spix found it near Rio de Janeiro in Brazil. It utters a note of alarm very like that of the common English thrush, (*Turdus musicus*).

2. Turdus Falklandicus. *Quoy et Gaim.*

T. Falklandicus, *Quoy et Gaim.* Zool. de l'Uranie, p. 104.
———— *Pernetty,* Hist. d'un Voy. aux Iles Malouines, II. p. 20.
———— *D'Orb. & Lafr.*, Voy. de l'Amer. Mer. Av. p. 202.
T. Magellanicus, *King,* Proc. Zool. Soc. (1830) p. 14.
———— *D'Orb. & Lafr.* Mag. de Zool. 1835. p. 16.

M. D'Orbigny has pointed out that the *Turdus Magellanicus* of King is only the male bird of *Turdus Falklandicus*. I obtained specimens from the Rio Negro, Falkland Islands, Tierra del Fuego and Chiloe: I believe I saw the same species in the valleys of Northern Chile ; I was informed that the thrush there lines its nest with mud, in which respect it follows the habits of species of the northern hemisphere. In the Falkland Islands it chiefly inhabits the more rocky and dryer hills. It haunts also the neighbourhood of the settlement, and very frequently may be seen within old sheds. In this respect, and generally in its habits, it resembles the English thrush (*Turdus musicus*): its cry, however, is different. It is tame, silent, and inquisitive.

1. Mimus Orpheus. *G. R. Gray.*

Orpheus Calandria, *D'Orb. & Lafr.* Mag. de Zool. (1835) p. 17.—Voy. de l'Amer. Mer. Av. 206. pl. x. f. 2.
Turdus Orpheus, *Spix.* Av. t. 1. pl. 71.
Mimus saturninus, *P. Max.* Beitr. p. 658 ?
Orpheus modulator, *Gould,* in Proc. of Zool. Soc. Part IV. (1836) p. 6.

This bird is described in the Proceedings of the Zoological Society (Part IV. 1836, p. 6.) as having come from the Straits of Magellan, which undoubtedly is a mistake. It is extremely common on the banks of the Plata; but a few degrees south of it, is replaced by the *O. Patagonica* of D'Orbigny. In Banda Oriental these birds are tame and bold; they constantly frequent the neighbourhood of the country houses to pick the meat, which is generally suspended to the posts and walls. If any other small bird joins in the feast, the Calandria (as this species is usually called in La Plata) immediately chases him away. In these respects, and in its manner of sometimes catching insects, the Mimus is related in its habits with that division of the *Muscicapidæ*, which includes the genus *Xolmis :* indeed, the general colour of the plumage of *X. Nengeta* is so like that of Mimus, that it might readily be mistaken for a bird of that genus. The Calandria haunts thickets and hedge-rows, where it actively hops about, and in doing so often elevates and slightly expands its tail.

2. Mimus Patagonicus. *G. R. Gray.*

Orpheus Patagonicus, *D'Orb. & Lafr.* Mag. de Zool. 1836, p. 19.—Voy. de l'Amer. Mer. Av. p. 210, pl. xi. f. 2.

I obtained specimens of this bird at the Rio Negro and at Santa Cruz in Southern Patagonia, at both of which places it is common. It is not found in Tierra del Fuego, for neither it nor the other species of the genus inhabit forests. This species has slightly different habits from the *M. Orpheus*. It is a shyer bird, and frequents the plains and valleys thinly scattered with stunted and thorn-bearing trees. It does not appear to move its tail so much. Its cry, like that of the rest of the genus, is harsh, but its song is sweet. The *M. Patagonicus*, whilst seated on the highest twig of some low bush, often enlivens the dreariness of the surrounding deserts by its varying song. Molina, however, describing the song of an allied species, has greatly exaggerated its charms. It may be compared to that of the sedge-bird (*Motacilla salicaria,* Linn.), but is much more powerful, some harsh notes and some very high ones being mingled with a pleasant warbling. The song of the different mocking thrushes certainly is superior to that of any other bird which I heard in South America; and they are almost the only ones which formally perch themselves on an elevated twig for the purpose of singing. They sing only during the spring of the year. I may here mention, as a curious instance of the fine shades of difference in habits between very closely allied species, that when I first saw the *M. Patagonicus*, I concluded from habits alone that it was different from *M. Orpheus*. But having afterwards procured a specimen of the former, and comparing the two without particular care, they appeared so very similar that I changed my opinion. Mr. Gould, however, immediately upon seeing them (and he did not then know that M. D'Orbigny had described them as different) pronounced that they were distinct species; a conclusion in conformity with the trifling difference of habit and geographical range, of which he was not at the time aware.

3. Mimus Thenca. *G. R. Gray*.

Turdus Thenca. *Mol.*
Orpheus Thenca. *D'Orb.* Voy. de l'Amer. Mer. Orn. p. 209, pl. f. 3.

This species seems to be confined to the coast of the Pacific, west of the Cordillera, where it replaces the *M. Orpheus*, and *M. Patagonicus* of the Atlantic side of the continent. Its southern limit is the neighbourhood of Concepcion, (lat. 37° S.) where the country changes from thick forests to an open land. The Thenca, (which is the name of this species, in the language of the Aboriginal Indians,) is common in central and northern Chile, and is likewise found (I believe the same species) near Lima, (lat. 12°) on the coast of Peru. The habits of the Thenca are similar, as far as I could perceive, to those of the *M. Patagonicus*. I observed many individuals, which had their heads stained yellow from the pollen of some flower, into which they bury their heads, probably for the sake of the small beetles concealed there. Molina describes the nest of the Thenca, as having a long passage, but I was assured by the country people, that this nest belonged to the *Synallaxis ægithaloides*, and that the Thenca makes a simple nest, built externally of small prickly branches of the mimosa.

4. MIMUS TRIFASCIATUS. *G. R. Gray.*

PLATE XVI.

Orpheus trifasciatus. *Gould*, in Proc. of Zool. Soc. Part v. 1837, p. 27.

M. vertice, nuchâ, et dorso nigrescentibus; uropygio rufo pallidè lavato; alis nigrescentibus, tectricibus notâ albescente terminali fascias tres transversas facientibus; rectricibus caudæ duabus intermediis nigrescentibus, reliquis ad apicem pallidioribus; plumis auricularibus, strigâ superciliari, gulâ, et corpore subtùs albis, lateribus notis guttisque fuscis ornatis; rostro pedibusque nigris.

Long. tot. 10⅝ unc.; rost. 1⅜; alæ, 5; caudæ, 5½; tarsi, 1¾.

The vertex, nape of the neck and the back, blackish; with the lower part of the back tinged with pale rufous; the wings blackish, with the tips of the wing coverts white, forming three transverse bands; the tail with the two intermediate feathers black, with the tips of the others much paler; the auricular feathers with a streak above the eyes, throat, and beneath the abdomen white; the flanks ornamented with fuscous marks and spots.

Habitat, Charles Island, Galapagos Archipelago. (*October*).

5. MIMUS MELANOTIS. *G. R. Gray.*

PLATE XVII.

Orpheus melanotis, *Gould*, in Proc. of Zool. Soc. Part v. 1837, p. 27.

M. vertice, nuchâ, dorsoque pallidè fuscis; plumis capitis et dorsi ad medium colore saturatiore; alis intensè fuscis, singulis plumis ad marginem pallidioribus, secundariis, tectricibusque majoribus notâ albâ terminali, fascias duas transversas facientibus; caudæ rectricibus nigrescenti-fuscis ad apicem albis, loro, plumisque auricularibus nigrescenti-fuscis; laterum plumis notâ fuscâ centrali, abdomine albo; rostro pedibusque nigris.

Long tot. 9½ unc.; rost. 1¼; alæ, 4½; caudæ, 4½; tarsi, 1⅜.

The vertex, nape of the neck and the back, pale brown; the feathers of the head and the back, as far as the middle, of a darker colour; the wings intensely brown, with the margins of each of the feathers paler; the secondaries and the greater wing-coverts terminated with white marks, giving the appearance of two transverse bands; the feathers of the tail blackish brown, with the tips white; the lores and the feathers of the ears blackish brown, the feathers of the sides with a central brown mark, the abdomen white; the bill and feet black.

Habitat, Chatham and James's Islands, Galapagos Archipelago. (*October.*)

6. MIMUS PARVULUS. *G. R. Gray.*
PLATE XVIII.

Orpheus parvulus. Gould, in Proc. of Zool. Soc. Part v. 1837, p. 27.

M. vertice, nuchâ caudâque intensè fuscis, hujus rectricibus ad apicem albo notatis; alis fuscis secundariis tectricibusque notâ albâ apicali fascias duas transversas facientibus; loro plumisque auricularibus nigrescentibus; gulâ, colli lateribus, pectore, et abdomine albescentibus; plumis laterum notis fuscis per medium longitudinaliter excurrentibus.

Long. tot. 8¼ unc.; *rost.* 1; *alæ*, 3⅜; *caudæ*, 3¾; *tarsi*, 1¼.

The vertex, the nape of the neck, and the tail intensely black; with the tips of the tail feathers marked with white; the wings brown with the secondaries and coverts tipped with white marks, giving the appearance of two transverse bands; the lores and the feathers of the ears black; the throat, the sides of the neck, breast, and the abdomen white; the flanks marked longitudinally with brown.

Habitat, Albemarle Island, Galapagos Archipelago. (*October.*)

It will be seen, that the three last species of the genus Mimus, were procured from the Galapagos Archipelago; and as there is a fact, connected with their geographical distribution, which appears to me of the highest interest, I have had these three figured. There are five large islands in this Archipelago, and several smaller ones. I fortunately happened to observe, that the specimens which I collected in the two first islands we visited, differed from each other, and this made me pay particular attention to their collection. I found that all in Charles Island belonged to *M. trifasciatus*; all in Albemarle Island to *M. parvulus*, and all in Chatham and James's Islands to *M. melanotus*. I do not rest this fact solely on my own observation, but several specimens were brought home in the Beagle, and they were found, according to their species, to have come from the islands as above named. Charles Island is distant fifty miles from Chatham Island, and thirty-two from Albemarle Island. This latter is only ten miles from James Island, yet the many specimens procured from both belonged respectively to different species. James and Chatham, which possess the same species, are seventy miles apart, but Indefatigable Island is situated between them, which perhaps, has afforded a means of communication. The fact, that islands in sight of each other, should thus possess peculiar species, would be scarcely credible, if it were not supported by some others of an analogous nature, which I have mentioned in my Journal of the Voyage of the Beagle. I may observe, that as some naturalists may be inclined to attribute these differences to local varieties; that if birds so different as *O. trifasciatus*, and

O. parvulus, can be considered as varieties of one species, then the experience of all the best ornithologists must be given up, and whole genera must be blended into one species. I cannot myself doubt that *M. trifasciatus*, and *M. parvulus* are as distinct species as any that can be named in one restricted genus.

The habits of these three species are similar, and they evidently replace each other in the natural economy of the different islands; nor can I point out any difference between their habits and those of *M. Thenca* of Chile; I imagined, however, that the tone of their voice was slightly different. They are lively, inquisitive, active birds, and *run* fast; (I cannot assert, positively, that *M. Thenca runs*). They are so extremely tame, a character in common with the other birds of this Archipelago, that one alighted on a cup of water which I held in my hand, and drank out of it. They sing pleasantly; their nest is said to be simple and open. They seem to prefer the dry sterile regions nearer the coast, but they are likewise found in the higher, damper and more fertile parts of the islands. To these latter situations, however, they seem chiefly attracted by the houses and cleared ground of the colonists. I repeatedly saw the *M. melanotis* at James Island, tearing bits of meat from the flesh of the tortoise, which was cut into strips and suspended to dry, precisely in the same manner as I have so often observed the *M. Orpheus*, in La Plata, attacking the meat hung up near the Estancias.

1. Furnarius rufus. *Vieill.*

Furnarius rufus, *Vieill.*, Ency. Meth. 513.
Merops rufus, *Gmel.* Pl. enl. 739.
Opetiorhynchus rufus, *Tem.* Man.
Turdus vadius, *Licht.* Cat.
Figulus albogularis, *Spix.* Av. pl. lxxviii. f. 1 & 2.
Fournier, *Buff., Azara*, No. 221.

This bird is common in Banda Oriental, on the banks of the Plata; but I did not see it further southward. It is called by the Spaniards Casaro, or house-builder, from the very singular nest which it constructs. The most exposed situation, as on the top of a post, the stem of an opuntia, or bare rock, is chosen. The nest consists of mud and bits of straw; it is very strong, and the sides are thick; in shape it resembles a depressed beehive or oven, and hence the name of the genus. Directly in front of the mouth of the nest, which is large and arched, there is a partition, which reaches nearly to the roof, thus forming a passage or ante-chamber to the true nest. At Maldonado, in the end of May, the bird was busy in building. The Furnarius is very common in Banda Oriental; it often haunts the bushes in the neighbourhood of houses; it is an active bird, and both *walks* and *runs* quickly, and generally by starts; it feeds chiefly on Coleoptera; it often utters a peculiar, loud, shrill, and quickly reiterated cry.

2. Furnarius cunicularius. *G. R. Gray.*

Alauda cunicularia, *Vieill.*
Alauda fissirostra, *Kittl.* Mem. l'Acad. St. Peters. ii. pl. 3.
Certhilauda cunicularia, *D'Orb. & Lafr.* Mag. de Zool.

This bird has a considerable geographical range. On the eastern side of the continent it is found from about 40° (for I never saw one in the southern districts of Patagonia) northward to at least 30°, and perhaps much further. On the western side its southern limit is the neighbourhood of Concepcion, where the country becomes dry and open, and it ranges throughout Chile (specimens were procured from Valparaiso) to at least as far north as Lima, in lat. 12°, on the coast of Peru. I may here observe, that the northern limit of all birds, which are lovers of dry countries, such as this Furnarius and some of the species of Mimus, is not probably at Lima but near Cape Blanco, 10° south of the Equator, where the open and parched land of Peru blends (as it was described to me) rather suddenly into the magnificent forests of Guayaquil. This Furnarius constantly haunts the driest and most open districts; and hence sand-dunes near the coast afford it a favourite resort. In La Plata, in Northern Patagonia, and in Central Chile, it is abundant: in the former country it is called Casarita, a name which has evidently been given from its relationship with the Casaro, or *Furnarius rufus*, for, as we shall see, its nidification is very different. It is a very tame, most quiet, solitary little bird, and like the English robin (*Sylvia rubecula*) it is usually most active early in the morning and late in the evening. When disturbed it flies only to a short distance; it is fond of dusting itself on the roads; it walks and runs (but not very quickly), and generally by starts. I opened the stomachs of some, and found in them remains of Coleoptera, and chiefly Carabidæ. At certain seasons it frequently utters a peculiar, shrill but gentle, reiterated cry, which is so quickly repeated as to produce one running sound. In this respect, and in its manner of walking on the ground, and in its food, this species closely resembles the Casaro, but in its quiet manners it differs widely from that active bird. Its nidification is likewise different, for it builds its nest at the bottom of a narrow cylindrical hole, which is said to extend horizontally to nearly six feet under ground. Several of the country people told me, that when boys, they had attempted to dig out the nest, but had scarcely ever succeeded in getting to the end. The bird chooses any low bank of firm sandy soil by the side of a road or stream. At the settlement of Bahia Blanca the walls are built of hardened mud; and I noticed one, enclosing a courtyard, where I lodged, which was penetrated by round holes in a score of places. On asking the owner the cause of this, he bitterly complained of the little Casarita, several of which I afterwards observed at work. It is rather curious, that as these birds were constantly flitting backwards and forwards over the low wall, they must be quite incapable of judging of distance or thickness even after the shortest circuitous route, for otherwise they would not have made so many vain attempts.

UPPUCERTHIA DUMETORIA. *I. Geoffr. & D'Orb.*

PLATE XIX.

Uppucerthia dumetoria, *J. Geoffr. & D'Orb.* Ann. du Mus. i. 393 and 394.
Furnarius dumetorum, *D'Orb.* MS.
Uppucerthia dumetorum, *D'Orb. & Lafr.* Mag. de Zool. 1838, p. 20.

This bird is an inhabitant of extremely sterile regions. I saw several at the Rio Negro, but at Port Desire they were, perhaps, more numerous. I did not observe it near Valparaiso, in Central Chile, but I procured specimens of it from Coquimbo, where the country is more desert. It frequents open places, in which a few bushes grow. It hops very quickly, and often flies quietly from one place to another. It may often be seen turning over and picking dry pieces of dung. It is a remarkable circumstance, that in the three specimens which I brought home, from different localities, namely the Rio Negro, Port Desire, and Coquimbo, the beak varies considerably in length: in that from Port Desire in Patagonia it is three-eighths of an inch shorter than in that from Coquimbo in Chile; whilst the Rio Negro specimen is intermediate between them. Mr. G. R. Gray has pointed out to me that Latham long since observed a great variation in the beak of the Patagonian warbler, *Opetiorhynchus Patagonicus.*

1. OPETIORHYNCHUS VULGARIS. *G. R. Gray.*

Uppucerthia vulgaris, *D'Orbig. & Lafr.* Mag. de Zool. 1838, p. 23.

This bird in general habits has several points of resemblance with the *Furnarius cunicularius,* but differs in some other respects. Its flight is somewhat similar, but it shows two red bands on its wings, instead of one, by which it can be distinguished at a distance: instead of walking it only hops; it feeds entirely on the ground, and in its stomach I found scarcely anything but Coleopterous insects, and of these many were fungi feeders. It often frequents the borders of lakes, where the water has thrown up leaves and other refuse. It likewise may be met with in all parts of the open grassy plains of Banda Oriental, where (like the *Uppucerthia* at the Rio Negro) it often turns over dry dung. Its note is very like that of the *F. cunicularius,* but more acute, and consists of a shrill cry, quickly reiterated so as to make a running sound. I was informed that, like that bird, it builds its nest at the bottom of a deep burrow. This species is common in La Plata, the Falkland Islands, and Tierra del Fuego; in the latter it frequents the higher parts of the mountains, or those exposed to the western gales, which are free from forests, for it is a bird that exclusively lives in open countries and on the ground. I believe it is not found in Chile; nor is it common on the coast of Patagonia. This species in its habits is very different from the three following closely allied ones, since the latter never, or most rarely, leave the sea beach, whilst this bird, excepting by chance, is never seen there, but always in the interior country. Nevertheless with this marked difference in habits, (there are several other points beside that of the station frequented), if the preserved skins of *O. parvulus* and *O. vulgaris* were placed in the hands of any one, even perhaps of a practised ornithologist, he would at first hesitate to consider them distinct, although upon closer examination he would find many points of difference,—of which the much greater strength of the feet and the greater length of the tarsus are conspicuous in those species, which live amongst the stones on the sea beach.

2. Opetiorhynchus Patagonicus. *G. R. Gray.*

Patagonian Warbler, *Lath.* Syn. iv. p. 434.
Motacilla Patagonica, *Gmel.*
Motacilla Gracula, *Forst.* Draw. No. 100.
Sylvia Patagonica, *Lath.* Index, ii. 517.
Furnarius Lessonii, *Dumont.*
———— Chilensis, *Less.* Voy. de la Coqu. i. p. 671, n. Tr. d'Ornith. p. 307, pl. 75, f. 1.
Opetiorhynchus rupestris, *Kittl.* Mem. de l'Acad. St. Petersb. i. p. 188, pl. viii.
Uppucerthia rupestris, *D'Orb. & Lafr.* Mag. de Zool. 1838, p. 21.

This bird is extremely common on the sea shore of all the bays and channels of Tierra del Fuego; on the western coast it is replaced in Northern Chile by the *O. nigrofumosus*, and in the Falkland Islands by the *O. antarcticus*. As the habits of this species and those just named are quite similar, I shall describe them all together under the head of *O. nigrofumosus*. A specimen of *O. Patagonicus* from Chiloe has a bill rather more than two-tenths of an inch longer than in those from Tierra del Fuego; but as no other difference can be perceived, I cannot allow that this is a specific character any more than in the case of the *Uppucerthia*.

3. Opetiorhynchus antarcticus. *G. R. Gray.*

Certhia antarctica, *Garn.* Ann. des Sc. Nat. 1826.
Furnarius fuliginosus, *Less.* Voy. de la Coqu. Zool. i. p. 670.
Patagonian Warbler, *Lath.* ♀ in Dixon's Voy. App. No. 1, 359 and pl.

This species inhabits the Falkland Islands. My specimens were procured at the east island, from which, also, those described by the French naturalists came, and likewise that given in the Appendix to Dixon's Voyage. I have no doubt that it is peculiar to this group, for the foregoing species, which in the neighbouring mainland of Tierra del Fuego supplies its place and has precisely the same habits, has been examined by Mr. Gould and is considered distinct. The *O. antarcticus* has long been noticed by voyagers to the Falkland Islands from its extreme tameness: in the year 1763 Pernety states it was so tame that it would almost perch on his finger, and that in half an hour he killed ten with a wand.

4. OPETIORHYNCHUS NIGROFUMOSUS. *G. R. Gray*.

PLATE XX.

Uppucerthia nigrofumosa, *D'Orb. et Lafr.* Mag. de Zool. 1838, p. 23.
Opetiorhynchus lanceolatus, *Gould*, MS. and on plate XX.

My specimen was killed at Coquimbo, on the coast of Chile. It differs from *O. Patagonicus* in its larger size, much stronger feet and bill, and more dusky plumage, and in the white streak over the eye being less plainly marked. In this species the red band, which extends from the body obliquely across the wings in all the species, reaches to the third primary, whereas in *O. Patagonicus*, *O. vulgaris*, and *O. antarcticus*, that feather is not marked, or so faintly, as scarcely to be distinguishable. In the genus Furnarius, the wing feathers are marked in an analogous manner. I saw this species (as I believe) on the coast near the mouth of the valley of Copiapó.

I will now make a few remarks on the habits of these three coast species. The first, *O. antarcticus*, is confined, as I have every reason to believe, to the Falkland Islands. The second inhabits Tierra del Fuego, and in Chiloe and Central Chile is replaced by the local variety with a long beak, and this still further northward by the *O. nigrofumosus*. On the east side of the continent I do not believe these marine species extend so far northward. I never saw one on the shores of the Plata, but they occur in Central Patagonia. These birds live almost exclusively on the sea beach, whether formed of shingle or rock, and feed just above the surf on the matter thrown up by the waves. The pebbly beds of large rivers sometimes tempt a solitary pair to wander far from the coast. Thus at Santa Cruz I saw one at least one hundred miles inland, and I several times observed the same thing in Chile, which has likewise been remarked by Kittlitz, who has given a very faithful account of the habits of *O. Patagonicus*. I must add that I also saw this bird in the stony and arid valleys in the Cordillera, at a height of at least 3000 feet. In Tierra del Fuego I scarcely ever saw one twenty yards from the beach, and both there and at the Falkland Islands they may frequently be seen walking on the buoyant leaves of the *Fucus giganteus*, at some little distance from the shore. In these respects, the birds of this genus entirely replace in habits many species of Tringa. In the stomachs of those I opened I found small crabs and little shells, and one Buccinum even a quarter of an inch long: Kittlitz says, he found in one, besides such objects, some small seeds. They are very quiet, tame and solitary, but they may not unfrequently be seen in pairs. They hop and likewise *run* quickly; in which latter respect, and likewise in their greater tameness, they differ from the *O. vulgaris*. Their cry is seldom uttered, but is a quick repetition of a shrill note, like that of the last named bird, and of several species of Furnarius.

On the 20th of September, I found, near Valparaiso, the nest of *O. Patagonicus*, with young birds in it: it was placed in a small hole in the roof of a deep cavern, not far from the bank of a pebbly stream. Three months later in the summer I found, in the Chonos Archipelago (Lat. 45°), a nest of this species, placed in a small hole beneath an old tree, close to the sea-beach. The nest was composed of coarse grass and was untidily built. The egg rather elongated; length 1·11 of an inch, width in broadest part ·8 of an inch; perfectly white.

Genus.—Eremobius. *Gould.*

Rostrum *capitis longitudine seu longius, fere rectum, ad apicem deorsum curvatum, haud emarginatum; naribus parvis, basalibus, oblongis, in sulco positis;* Alæ *breves, remigibus primariis secundariisque fere æqualibus, plumis 4, 5, 6-que subæqualibus longissimisque;* Cauda *mediocris apice rotundato;* Tarsi *sublongi antice squamis ferè obsoletis induti, halluce digito medio breviore, digitis lateralibus inæqualibus, internis brevioribus.*

Eremobius phœnicurus. *Gould.*

Plate XXI.

E. fuscus, remigibus cinereo fusco marginatis, striâ superciliari pone oculos extensâ cinereo-albâ; caudâ nigro-fuscâ basi castaneo fuscâ; gulâ abdomineque medio cinereo albis; hypochondriis tectricibusque caudalibus inferioribus pallide flavescentibus.

Long. tot. 6$\frac{9}{12}$ unc.; *rost.* 1; *alæ*, 2$\frac{9}{12}$; *caudæ*, 3; *tarsi*, $\frac{9}{12}$.

Head and all the upper surface brown; the primaries margined with greyish brown; stripe over and behind the eye greyish white; tail feathers chestnut brown at the base, and blackish brown for the remainder of their length;

throat and centre of the abdomen greyish white, passing into pale buff on the flanks and under tail-coverts; bill and feet blackish brown.

Habitat, Patagonia.

This bird, though forming a well-marked genus, is in many respects, even in plumage, allied to Furnarius and Opetiorhynchus,—for instance, in the streak over its eyes, in the red band on its wings extending obliquely from the body to the third primary, and to some of the species of these genera in its rather plumose feathers. In its general manners, the same resemblance, together with some differences, always struck me. It lives entirely on the ground, and generally in dry sterile situations, where it haunts the scattered thickets, and often flies from one to another. When skulking about the bushes it cocks up its tail, imitating in this respect Pteroptochos and Rhinomya. Its cry is shrill, quickly reiterated, and very similar to that of several species of Furnarius and Opetiorhynchus. The stomach of one which I opened was full of Coleoptera. I procured specimens from three places on the coast of Patagonia; namely, Port Desire, St. Julian, and Santa Cruz; but it is nowhere common. I likewise saw it at a considerable elevation in the eastern valleys of the barren Cordillera, near Mendoza.

Rhinomya lanceolata. *Is. Geoffr. & D'Orb.*

Rhinomya lanceolata. *Is. Geoffr. & D'Orb.* Voy. de l'Amer. Mer. pl. 7. f. 1. 1832, cl. 11. pl. 3. id.—Mag. de Zool. 1832, 11. pl. 3. and 1837, p. 15.

I procured a specimen of this bird from the Rio Negro in Northern Patagonia, and I never saw one any where else; and M. D'Orbigny makes the same remark. On the Atlantic side of the continent, it replaces the several species of *Pteroptochos* which live on the shores of the Pacific. Its habits, in some respects, are similar; it lives at the bottom of hedges or thickets, where it runs with such quickness, that it might easily be mistaken for a rat. It is very unwilling to take flight, so that, I was assured by some of the inhabitants, that it could not fly, which, however, is a mistake. It frequently utters a loud and very singular cry. The Rhinomya is distantly allied to the *Eremobius phœnicurus*, which is found in Southern Patagonia, whose habits in some respects are similar.

1. Pteroptochos Tarnii. *G. R. Gray.*

Hylactes Tarnii. *Vigors*, Proc. Zool. 1830.
Megalonyx ruficeps. *D'Orb. & Lafr.* Mag. de Zool. 1837. p. 15.
Leptonyx Tarnii. *D'Orb. & Lafr.* Voy. de l'Amer. Mer. Av. p. 198, pl. viii. f. 1.

This species, as well as several others of the genus, and likewise of Scytalopus are confined to the west coast of South America. The *P. Tarnii* ranges from the neighbourhood of Concepcion, lat. 37°, to south of the Peninsula of Tres Montes, between 41° and 50°. It is not found in Tierra del Fuego, where the climate probably is too cold for it, for in other respects, the great forests of that country appear admirably adapted to its habits. Its limit, northward of the province of Concepcion, is evidently due to the change which there takes place, from dense forests to an open and dry country. The *P. Tarnii* is abundant in all parts of the Island of Chiloe, where it is called by the native Indians, *guid-guid;* but by the English sailors, the barking-bird. This latter name is very well applied, for the noise which it utters is precisely like the yelping of a small dog. When a person is walking along a pathway within the forest, or on the sea-beach, he will often be surprised to hear on a sudden, close by him, the barking of the *guid-guid*. He may often watch in vain the thicket, whence the sound proceeds, in hopes of seeing its author, and if he endeavour, by beating the bushes, to drive it out, his chance of success will be still smaller. At other times, by standing quietly within the forest, the *guid-guid* will fearlessly hop close to him, and will stand on the trunk of some dead tree, with its tail erect, and strange figure full in view. It feeds exclusively on the ground, in the thickest and most entangled parts of the forest. It rarely takes wing, and then only for short distances. It has the power of hopping quickly and with great vigour; when thus awkwardly proceeding, it carries its short tail in a nearly erect position. I was informed that the *guid-guid*, builds a nest amongst rotten sticks, close to the ground.

2. Pteroptochos megapodius. *Kittl.*

Pteroptochos megapodius. *Kittl.* 1830, Mem. de l'Acad. 1, pl. iv. et Vogel. von Chili, p. 10, pl. iv.
Megalonyx rufus. *Less.* Cent. Zool. 1831, pl. 66.
————— *D'Orb. & Lafr.*
Leptonyx macropus. *Swains.* Zool. Ill. pl. 117.
————— *D'Orb. & Lafr.* Voy. de l'Amer. Mer. Av. 197.

This bird is common in the dry country of central and northern Chile, where it replaces the *P. Tarnii* of the thickly wooded southern regions. The *P. megapodius,* is called by the Chilenos, " *El Turco* ;" it lives on the ground amongst the bushes which are sparingly scattered over the stony hills. With its tail erect, every now and then it may be seen popping on its stilt-like legs from one bush to another with uncommon celerity. Its appearance is very strange and almost ludicrous, and the bird seems always anxious to hide itself. It does not run, but hops, and can hardly be compelled to take flight. The various loud cries which it utters, when concealed in the bushes, are as strange as its appearance. I opened the extremely muscular gizzards of several of these birds, and found them filled with beetles, vegetable fibres, and pebbles. Observing the structure of the gizzard, the fleshy covering to the nostrils, and the arched, rounded wing, and great scratching claws, it was easy to imagine some distant kind of relationship between these birds and those of the Gallinaceous order. I was informed that the Turco makes its nest at the bottom of a deep burrow which it excavates in the ground.

3. PTEROPTOCHOS ALBICOLLIS. *Kittl.*

Pteroptochos albicollis. *Kittl.* Mem. de l'Acad. Petersb. 1. pl. iii. Vogel von Chili; p. 8. pl. iii.
Megalonyx medius. *Less.* Ill. Zool. pl. lx.
Megalonyx albicollis. *D'Orb. and Lafr.* Mag. de Zool. (1636,) Aves, p. 15.
Leptonyx albicollis. *D'Orb.* Voy. de l'Amer. Mer. Av. p. 196, pl. viii. f. 2.

This species is called by the Chilenos "Tapacolo," or cover your posteriors. The name is well applied, as the Tapacolo generally carries its short tail more than erect, that is, inclined backward and toward the head. It is extremely common in central Chile; and in the same manner as the Turco replaces the Barking-bird of the southern forest-land, so does the Tapacolo replace a fourth species (*P. rubecula*), which is an inhabitant of the same forests. The Tapacolo frequents hedge-rows, and the bushes which are scattered at a considerable elevation over the sterile hills, where scarcely another bird can exist: hence it plays a conspicuous part in the ornithology of Chile. In its manner of feeding, and quickly hopping out of a thicket and back again, in its desire of concealment, unwillingness to take flight, and nidification, it manifests a close resemblance with the *P. megapodius*; its appearance is not, however, so strange, and (as if in consequence) it exposes itself more readily to view. The Tapacolo is very crafty; when frightened by any person, it will remain motionless at the bottom of a bush, and will then, after a little while, try with much address to crawl away on the opposite side. It is also an active bird, and continually making a noise; these noises are various and strangely odd; one is like the cooing of doves, another like the bubbling of water, and many defy all similes. The country people say it changes its cry five times in the year, which is according, I suppose, to some change of season. I was told that the Tapacolo builds its nest at the bottom of a deep burrow, like the Turco; whereas the *P. Tarnii*, (as well as the *P. rubecula*, an inhabitant of the same districts,) makes its nest amongst the sticks just above the ground. This difference in the nidification, of the southern and northern species, is probably due to the nature of the damp forests inhabited by the former in which a burrow could hardly be made dry. I may here observe, that travelling northward from Valparaiso to Coquimbo, I met near Illapel with a bird closely allied to the Tapacolo, but which, from some slight difference in manners, I believed was a distinct species. The range of this supposed species, is from between Coquimbo and Valparaiso, to at least as far north as the valley of Copiapó.

4. Pteroptochos rubecula. *Kittl.*

Pteroptochos rubecula, *Kittl.* Vog. von Chili, p. 7. pl. ii.
Megalonyx rubecula, *D'Orb. & Lafr.* Mag. de Zool. 1837, p. 16.
Megalonyx rufogularis, *D'Orb. & Lafr.* Voy. de l'Amer. Mer. pl. 7, f. 2.
Leptonyx rubecula, *D'Orb. & Lafr.* Voy. de l'Amer. Mer. Av. p. 196.

This species appears to have nearly the same range with the *P. Tarnii*: its southern limit certainly extends as far as 47° south, but northward, where the forests cease, near Concepcion, I was unable to ascertain that this bird is ever met with, and Kittlitz has made the same remark. In Chiloe, where it is common, it is called by the Indian inhabitants the "Cheucau." It frequents the most gloomy and retired spots within the damp forests. Sometimes, although the cry of the Cheucau is heard close by, a person may watch attentively and yet in vain; at other times, if he stands motionless, the red-breasted little bird will approach within a few feet, in the most familiar manner. It then busily hops about the entangled mass of rotting canes and branches, with its little tail cocked upwards. I opened the gizzard of several specimens; it was very muscular, and contained hard seeds, buds of plants, occasionally some insects, and vegetable fibres mixed with small stones. The Cheucau is held in superstitious fear by the Chilotans, on account of its strange and varied cries. There are three very distinct kinds:—one is called "chiduco," and is an omen of good; another "huitreu," which is extremely unfavourable; and a third, which I have forgotten. These words are given in imitation of its cries, and the natives are in some things absolutely governed by them. I have already stated that I was informed by the inhabitants that the Cheucau builds its nest amongst sticks close to the ground.

5. Pteroptochos paradoxus. *G. R. Gray.*

Troglodytes paradoxus, *Kittl.* Vog. von Chili, p. 12, pl. 5.—*Id.* Mem. de l'Acad. St. Peters. 1833, i. pl. 5.
Malacorhynchus Chilensis, *Kittl.* Mem. de l'Acad. St. Peters. 1835, p. 527.
Leptonyx paradoxus, *D'Orb.* Voy. de l'Amer. Mer. Av. p. 197.

This species differs in a small degree from all the others of the genus: its claws are longer, tarsi shorter, and bill flattened at the top: in these, and some other respects, it approaches to Scytalopus. I may add, that from a greater degree of resemblance, especially in the feet, *P. Tarnii* and *megapodius* may be ranked in one section, and *P. albicollis* and *rubecula* in another.

I procured specimens of the *P. paradoxus* both from Valdivia and Chiloe; like the *P. Tarnii* and *P. rubecula* it is confined to the regions of forest. Its habits are closely similar to those of the last species. I opened the gizzard of one at Valdivia, and found it full of large seeds and the remnants of insects. In

Chiloe, where it is much less common than the Cheucau, it is called by the inhabitants Cheuqui. Kittlitz procured specimens from Concepcion. He describes the cry which it utters over and over again, in the same high tone, as very singular, and more like that of a frog than of a bird.

Scytalopus Magellanicus. *G. R. Gray.*

Sylvia Magellanica, *Lath.* Index, ii. p. 528. ♀ Forst. Dr. No. 163. ♀
Scytalopus fuscus, *Gould*, in Proc. of Zool. Soc. Part iv. 1836, p. 39. ♂
——————— *Jard. and Selb.* Ill. Orn. New Ser. pl. 19. ♂
Platyurus niger, *Swains.*, Two Cent. and a Quarter, p. 323. ♂

This bird has a wider range than the species of the foregoing and closely allied genus. It is common near Port Famine in Tierra del Fuego, and on the west coast in the thickly wooded islets of the Chonos Archipelago. I was assured by an intelligent collector that this bird is met with, though rarely, in central Chile; and Mr. Gould informs me, that he has received specimens from that country. It has found its way over to the Falkland Islands, where, instead of inhabiting forests, it frequents the coarse herbage and low bushes, which in most parts conceal the peaty surface of that island. In general appearance the *Scytalopus fuscus* might at first be mistaken for a Troglodytes, but in habits it is closely allied to the several species of Pteroptochos. In a skulking manner, with its little tail erect, it hops about the most entangled parts of the forests of Tierra del Fuego; but when near the outskirts, it every now and then pops out, and then quickly back again. It utters many loud and strange cries: to obtain a good view of it is not always easy, and still less so to make it fly. A specimen I procured at Chiloe had its upper mandible stronger and more arched, but differed in no other respect.

1. Troglodytes Magellanicus. *Gould.*

T. Magellanicus, *Gould*, in Proc. of Zool. Soc. Part iv. 1836, p. 88.

This bird has a considerable range. I procured specimens of it near Rio de Janeiro, on the banks of the Plata, throughout Patagonia, in Tierra del Fuego, where it is one of the commonest birds, and likewise in Central Chile: its habits resemble very closely those of the common Troglodytes of England. In the open country near Bahia Blanca it lived amongst the thickets and coarse herbage in the valleys; in Tierra del Fuego, in the outskirts of the forest. Its chirp is harsh. In Chile I saw one in October building its nest in a hole in a stone wall, in a situation such as would have been chosen by our Troglodytes.

2. Troglodytes platensis. *Gmel.*

I procured specimens of this bird from Bahia Blanca, in Northern Patagonia, and likewise from the Falkland Islands, where it is not uncommon. When first killed, its legs and beak appear of larger size, compared to its body, than in other species of this genus. In the Falkland Islands it lives, almost exclusively, close to the ground, in the coarse grass which springs from the peaty soil. I do not think I ever saw a bird which, when it chose to remain concealed, was so difficult to disturb. I have frequently marked one down to within a yard on the open grassy plain, and afterwards have endeavoured, quite in vain, by walking backwards and forwards, over the same spot, to obtain another sight of it.

1. Synallaxis humicola. *Kittl.*

S. humicola, *Kittl.* Mem. de l'Acad. St. Peters. i. pl. 6.—*Id.* Vog. von Chili, p. 13, pl. vi.

Not uncommon in the neighbourhood of Valparaiso. Kittlitz has well described its habits. He says it lives on the ground under thickets, that it is active in running about, and that it readily flies from bush to bush. It holds its tail upright; utters a shrill, quickly reiterated cry; feeds on insects; but Kittlitz found in the stomach chiefly grains and berries, with little stones. From these circumstances, he conceives that this bird shews some affinity with Pteroptochos, but I feel no doubt that in the form of its beak, wings, tail, manner of carrying the latter, kind of plumage, sound of voice and habits, the relationship is much closer with Eremobius, which perhaps it may be considered as representing on the Pacific side of the Cordillera. Its tongue is furnished with bristly points, but apparently is less deeply bifid than in the other species of Synallaxis or Limnornis. I obtained both sexes, but there is no difference in their plumage.

For the reason just given, I have put this species at the head of its genus, and therefore nearest to Eremobius, although it is impossible to represent by a linear arrangement, the multiplied relations between the following genera— Furnarius, Uppucerthia, Opetiorhynchus, Eremobius, Anumbius, Synallaxis, Limnornis, Oxyurus; and again, Rhynomya, Pteroptochos, Scytalopus, and Troglodytes, which, with the exception of the last, are strictly South American forms.

2. SYNALLAXIS MAJOR. *Gould.*

PLATE XXII.

S. olivaceo fuscus; infra fulvus albo distinctè maculatus; plumis singulis stria obscura centrali notatis; fronte rufo, remigibus fuscis, cinereo-fusco externè maculatis, tertiariis nigro fuscis apice margineque latè cinereo-fuscis; gulâ albâ, plumarum flavescentium serie fusco maculatarum circumdatâ.

Long. tot. 8 unc.; *rost.* 1; *alæ*, 3¼; *caudæ*, 4; *tarsi*, 1.

Forehead rufous; crown of the head, back of the neck and back olive brown, with a conspicuous stripe of blackish brown down the centre of each feather; wing-coverts and lower part of the back olive brown, with a faint trace of the dark patch in the centre of each feather; primaries brown, margined externally with greyish brown; spurious wing and secondaries rufous tipped with brown; tertiaries blackish brown broadly margined and tipped with greyish brown; two centre tail feathers dark olive brown; the remainder blackish brown largely tipped with white; throat white encircled with a series of feathers of a buff colour spotted with dark brown; breast and all the under surface tawny indistinctly blotched with white; tarsi with a very pale blue tinge.

Habitat, Maldonado, north bank of La Plata. (*June*).

This bird is not very common. Those which I saw lived on the ground in dry and open places, and did not frequent the neighbourhood of lakes abounding with rushes or thickets, like the greater number of species of Synallaxis, and the allied genus Limnornis. The flight of this bird is peculiar, which seems chiefly due to the length of its elegantly acuminated tail. It sometimes alights and rests on the summit of a thistle or twig, a habit different from that of any species of the genus which I have seen. Its manner of living and feeding on the ground might have been suspected, from the length of the soft secondaries, like those of a lark or of *Furnarius cunicularius.* The claws also of the front toes are produced and perhaps they are rather straighter than in other members of the family. The tongue is bifid and divided into bristly points. The nest, of which I have seen two, is very peculiar. It is cylindrical, about two feet long, and placed vertically in the middle of a thick bush in an exposed situation. It is made externally of prickly branches, and is very large compared with the size of the bird. The opening is at the upper extremity, from which a passage leads to the true nest, which is lined with feathers and hairs. There is a slight bend in the passage both at its exit and where it enters the nest.

3. Synallaxis rufogularis. *Gould.*

Plate XXIII.

S. olivaceo fuscus plumis singulis maculâ oblongâ fusco nigrâ; remigibus primariis secundariisque basi ferrugineo fuscis, apice nigro fuscis, flavescenti albo marginatis; lineâ superciliari, mento abdomineque medio flavescenti albis; gulâ ferrugineo fuscâ; pectore fulvescenti fusco, plumis singulis striâ pallidiore centrali ornatis.

Long. tot. 6½ unc; *rost.* ⅝; *alœ*, 3; *caudœ*, 3¼; *tarsi*, 1.

Head and all the upper surface and two centre tail feathers, brown, with a large oblong patch of brownish black down the centre of each feather; primaries, except the three outer ones, bounded posteriorly with an irregular line of black; secondaries, rusty brown at the base, and brown for the remainder of their length, margined all round with greyish olive; lateral tail feathers brownish black, largely tipped with tawny white; stripe from the nostrils over each eye, chin, and centre of the abdomen, pale buff; sides of the face and throat grey, with a spot of dark brown down the centre of each feather; in the centre of the throat, a patch of ferruginous brown; chest, pale brownish buff, with a fine pale stripe down each feather; bill and feet brown.

Habitat, Patagonia. (*April.*) Valparaiso. (*September.*)

These birds are not uncommon on the dry rocky mountains near Valparaiso, and in the valleys of southern Patagonia, where a few thickets grow. They hop actively about the withered herbage and low thickets, and often feed on the ground. The hind claw is weaker and straighter than in most of the other species of this genus.

4. Synallaxis maluroides.

S. maluroides. *D'Orb. & Lafr.* Voy de l'Amer. Mer. Ois. pl. xiv, f. 2. Mag. de Zool. 1837, Cl. 11, pl. 22.

My specimens were shot near Maldonado. Iris yellow; tarsi very pale coloured.

This species, as well as some others of Synallaxis, Anumbius, and Limnornis, live amongst reeds and other aquatic plants on the borders of lakes, and have the same general habits. I will, therefore, here describe them. They all have the power of crawling very quickly by the aid of their powerful claws and feet, as I soon discovered when they were not killed at once, for then it was scarcely possible to catch them. Their soft tail-feathers show signs of being used, but they never apply them, as the Certhias do, as a means of supporting their bodies. The tail-feathers were (at least during June) so loosely attached, that I seldom procured a specimen with all of them perfect; and I saw many (especially of *S. maluroides*), flying about with no tail. All the species, or nearly all, utter an acute, but not loud, rapidly reiterated cry. They are active and busily seek for small insects, chiefly Coleoptera, in the coarse herbage. The iris in all is rusty red; the tongue is divided and terminates in bristly points. These reed birds, which are very numerous both in species and individuals, on the borders of lakes in the provinces north of the Plata, appear to supply in South America, the various Sylviæ, which frequent similar stations in Europe.

5. Synallaxis flavogularis. *Gould.*

Plate XXIV.

S. supra fuscescenti cinereus, infra cinereo-fuscus; remigibus obscurè fuscis, basi obscurè rufis; caudæ plumis sex mediis nigro-fuscis, externis ferrugineis; genis guláque flavescentibus, plumis singulis apice obscurè fuscis.

Long. tot. 6½ unc; *rost.* ⅜; *alæ*, 2½; *caudæ*, 3⅝; *tarsi*, ¾;

Head and all the upper surface, brown; primaries, dark brown, with the basal portions rufous; six central tail-feathers, blackish brown; the remainder ferruginous; sides of the face and throat yellowish, with the tip of each feather dark brown; the remainder of the under surface, greyish brown; bill and feet, dark brown.

Habitat, Patagonia.

My specimens were obtained at Bahia Blanca and at Santa Cruz, two extreme parts of Patagonia. It frequents the thinly scattered thickets on the arid plains: the hind claw of its foot is not produced as in *S. rufogularis*, and it lives less on the ground.

6. Synallaxis brunnea. *Gould.*

S. pallide rubro fusca; primariis secundariisque rufis apice fuscis; caudæ plumis quatuor mediis nigrescenti fuscis, duabus proximis ferrugineo fuscis internè nigrescenti-marginatis, duabus extimis ferrugineo fuscis; genis, gulá abdomineque medio albescentibus; hypochondriis cinereis.

Long. tot. 5$\frac{1}{12}$; unc. *rost.* $\frac{8}{12}$; *alæ*, 2$\frac{5}{12}$; *caudæ*, ⅝; *tarsi*, $\frac{10}{12}$.

Head and all the upper surface pale reddish brown; primaries and secondaries, brown at the tip and rufous at the base; four central tail feathers, blackish brown; the next on each side rusty brown, margined internally with blackish brown; the two lateral feathers wholly rusty brown; sides of the face, throat, and centre of the abdomen, whitish; flanks cinereous; bill and feet brown.

Habitat, Port Desire, Patagonia. (*January.*)

This little bird frequents the thickets in the dry valleys near Port Desire. It often flies from bush to bush, and its habits are nearly like those of the rest of the genus. From its tail feathers, however, being little used, and the tarsi being slightly elongated, I suppose it lives chiefly on the ground. I may observe, that this species comes nearest to *S. flavogularis*, but that in the form of its tail, straightness of bill, and kind of plumage, it departs from Synallaxis, and approaches Eremobius.

7. SYNALLAXIS ÆGITHALOIDES. *Kittl.*

S. Ægithaloides. *Kittl.* Mem. de l'Acad. 11. pl. vii.—Vog. von Chili, p. 15, pl. vii.

This bird is common throughout Patagonia and Central Chile, being found wherever thickets grow on a rocky or dry soil. It sometimes moves about in small flocks. Its habits, as Kittlitz remarks, resemble in many respects, those of a titmouse (*Parus*); but there is one remarkable point of difference, namely, that this bird is able to *run* very quickly on the ground. It does not always do so, but often hops about with great activity; nevertheless, I repeat, I have distinctly seen it running very quickly amongst the thickets. When hopping from twig to twig, it does not use its long tail, any more than the long-tailed titmouse (*Parus caudatus*) of Europe. It utters a harsh, shrill, quickly reiterated cry, like so many other species of this genus and the allied ones. In Chile, I several times saw a very large cylindrical nest, built of prickly twigs of the mimosa, and placed in the middle of a thorn-bearing bush, with its mouth at the upper extremity; I was assured by the country people, that although so very large, it belonged to this little bird.* This kind of nidification, the habit of feeding on the ground, and the length of acuminated tail, are points of resemblance with *S. major*.

8. SYNALLAXIS RUFICAPILLA. *Vieill.*

Synallaxis ruficapilla. *Vieill.* Gal. des Ois. pl. lxxiv.
Parulus ruficeps. *Spix.* Av. Sp. Nov. tom. 1. p. 84, t. lxxxvi. f. 1. m. f. 2. fem.
Sphenura ruficeps. *Licht.* Ver. p. 42.

My specimens were obtained at Maldonado, (June) where it was rare, and at Buenos Ayres. Near Santa Fè, in Entre Rios, 3° northward, it was common: Spix found it near the Rio San Francisco in Brazil. Iris yellowish red; legs with faint tinge of blue; tongue terminated in bristly points, not deeply bifid. This Synallaxis approaches in character *Anumbius ruber*. Habits similar to those of *S. maluroides*.

---

* Molina, in his account of Chile, attributes this nest, I believe, through an error, to *Mimus thenca*.

ANUMBIUS RUBER. *D'Orb. and Lafr.*

Anumbius ruber.   *D'Orb. & Laf.* Mag de Zool, 1838, p. 18.
Furnarius ruber.   *Vieill.* Ency. Meth. 514.
Anumbi rouge.   *Azara*, No. 220.

Frequents reeds on the borders of lakes near Maldonado. Habits very similar to those of *Synallaxis maluroides*, and likewise of the two species of Limnornis; to one of which *L. curvirostris*, it is most closely allied in structure. Iris bright yellowish orange; tarsi, with faint tinge of blue; tongue divided on each side a little below the extreme point.

Genus.—LIMNORNIS. *Gould.*

Rostrum *capitis longitudine seu longius, leviter a basi ad apicem arcuatum, lateraliter compressum, haud emarginatum; naribus magnis basalibus linearibus apertis aut partim operculo tectis:* alæ *brevissimæ rotundæ, plumis quarta, quinta sextaque ferè æqualibus et longissimis;* cauda *rotundata et graduata, scapis aliquanto ultra radios productis;* tarsi *mediocres, fortiter scutellati; halluce digito medio breviore, robusto, ungue robusto armato, digitis lateralibus ferè æqualibus, intermediis aliquantò brevioribus.*

1. Limnornis rectirostris. *Gould.*

Plate XXVI.

L. pallide flavescenti fusca; cervice nigrescenti fusco; caudâ rufâ; tectricibus primariis secundariisque fuscis rufo latè marginatis; fasciâ pone oculos, gulâ abdomineque flavescenti albis; hypochondriis fulvis.

Long. tot. 6$\frac{2}{12}$ unc; rost. $\frac{9}{12}$, alæ, 2$\frac{6}{12}$ caudæ, 2$\frac{1}{12}$ tarsi, $\frac{7}{12}$.

Crown of the head brown; the remainder of the upper surface, pale yellowish brown; tail rufous and acutely pointed; wing coverts, primaries and secondaries brown, broadly margined with rufous; stripe behind the eye, throat, and all the under surface buffy white; flanks tawny; bill lengthened, orange at the base, dark brown at the tip; iris rusty red; feet very pale coloured; claws whitish.

Habitat, Maldonado, La Plata. *(June.)*

This bird lives amongst the reeds on the borders of lakes. It often alights vertically on stems of plants, but in climbing does not use its tail: habits, generally similar to those of *Synallaxis maluroides*.

2. Limnornis curvirostris. *Gould.*

Plate XXV.

L. rufescenti-fusca; caudâ, remigiumque basibus pallidè castaneo-fuscis, lineâ superciliari, genis, gulâ abdomineque albis; hypochondriis cervino tinctis.

Long. tot. 7 unc., *rost.* 1⅛; *alæ*, 2₁⁷₂; *caudæ*, 3₁₁₂; *tarsi*, ¹⁹₁₂.

Head, all the upper surface, and wings reddish brown; tail and basal portion of the outer margins of the primaries and secondaries reddish chesnut brown; stripe over the eye, throat, and all the under surface white, tinged, especially on the flanks, with fawn colour; bill orange at the base, the tip brown; legs pale bluish; claws white; tongue bristled on the sides; near the extremity it is divided into little bristly points.

Habitat, Maldonado, La Plata. (*June.*)

This species frequents the same localities with the last, and I am unable to point out any difference in its habits. Of the two specimens collected, the beak of one is very nearly one-tenth of an inch longer than that of the other; but this is almost wholly due to the sharp point of the upper mandible projecting beyond the lower mandible in the one, whereas they are nearly equal in the other.

1. Oxyurus tupinieri. *Gould.*

Synallaxis tupinieri. *Less.* Zool. de la Coqu. pl. 29. f. 1.
Oxyurus ornatus, *Swains.* 2 Cent. and ¼. p. 324.

This bird is perhaps the most abundant of any land species inhabiting Tierra del Fuego. It is common along the west coast, (and numerous in Chiloe,) even as far north as a degree south of Valparaiso; but the dry country and stunted woods of central Chile are not favourable to its increase. In the dark forests of Tierra del Fuego, both high up and low down, in the most gloomy, wet, and scarcely penetrable ravines, this little bird may be met with. No doubt, it appears more common than it really is, from its habit of following, with seeming curiosity, every person who enters these silent woods; continually uttering a harsh twitter, it flutters from tree to tree, within a few feet of the intruder's face. It is far from wishing for the modest concealment of the creeper (*Certhia familiaris*); nor does it, like that bird, run up the trunks of trees, but industriously, after the manner of a willow wren, hops about and searches for insects on every twig and branch.

2. Oxyurus? dorso-maculatus. *Gould.*

Synallaxis dorso-maculata. *D'Orb. and Lafr.* Voy. de l'Amer. Mer. Ois. pl. 14. f. 1.
——————————————————— Mag. de Zool. 1837, Cl. 11. p. 21.

My specimen was procured from Maldonado, (*June*), where it was not common. It frequents the same localities with *Synallaxis maluroides*, and the two species of *Limnornis*, and has very similar habits with them. In structure, and in the general shade of its plumage, it is closely allied to the foregoing species, although differing from it in habits.

Genus.—DENDRODRAMUS. *Gould.*

Rostrum capitis longitudine, aut longius, culmine recto, gonide ascendente, per omnes partes lateraliter compressum, durum et apice inemarginatum, naribus basalibus longitudinalibusque; alæ mediocres et subacuminatæ, plumis tertia, quarta et quinta æqualibus longissimisque; cauda mediocris, scapis ultra radios in spinas acutas productis; tarsi sub-breves, digitis unguibusque longis, his multum curvatis, digito externo valido et ferè digiti medii longitudine, digitis lateralibus inæqualibus, internis multum brevioribus.

DENDRODRAMUS LEUCOSTERNUS. *Gould.*

Plate XXVII.

D. capite, dorsi parte superiore alisque nigrescenti fuscis, rubro-tinctis; primariis secundariisque subferrugineo fusco irregulariter marginatis, uropygio caudâque nitidè ferrugineis, gulâ pectoreque albis, abdomine medio rufescenti fusco, singulis plumis ad apicem maculâ magnâ ovali albâ; hypochondriis saturatè rufis; rostro basi corneo, apice pedibusque nigro fuscis.

Long. tot. 6$\frac{3}{12}$ unc.; *rostri*, 1$\frac{1}{12}$; *alæ*, 3; *caudæ*, 2$\frac{9}{12}$; *tarsi*, $\frac{9}{12}$.

Head, upper part of the back and wings blackish brown, tinged with red; primaries and secondaries irregularly margined with dull rusty brown; rump and tail rich ferruginous; throat and chest white; feathers of the centre of the abdomen reddish brown, with a large oval spot of white near the tip of each feather; flanks deep rufous; bill horny at the base, the remainder and the feet blackish brown.

Habitat, Chiloe and Southern Chile.

This bird is common in the forests of Chiloe, where, differently from the *Oxyurus tupinieri*, it may constantly be seen running up the trunks of the lofty forest trees. Its manners appeared to me to resemble those of *Certhia familiaris*. I found Coleopterous insects in its stomach. Its range does not appear to be extensive; Chiloe to the south, and some woods near Rancagua (a degree south of Valparaiso) were the extreme points where I met with it. The Dendrodramus is not found in Tierra del Fuego, where the *O. tupinieri* is so numerous. Mr. G. R. Gray remarks that this genus is very nearly allied to Dendroplex of Mr. Swainson.

Family.—SYLVIADÆ.

Sub-Fam.—MOTACILLINÆ.

1. Muscisaxicola mentalis. *D'Orb. & Lafr.*

M. mentalis, *D'Orb. & Lafr.* Mag. de Zool. 1837, p. 66.
———————— Voy. dans l'Amer. Mer. Ornith. pl. 40, f. 1.

I procured specimens of this bird from Bahia Blanca, in Northern Patagonia, from Tierra del Fuego, from Chiloe, and from Central and Northern Chile. It is everywhere common. It frequents open places; so that in the wooded countries it lives entirely on the sea-beaches, or near the summits of mountains, where trees do not grow. In the excessively sterile upper valleys of the Cordillera of Northern Chile I met with this bird, even at a height of little less than ten thousand feet, where the last traces of vegetation occur, and where no other bird lives. It generally moves about in very small flocks, and frequents rocky streams and marshy ground: it hops and flies from stone to stone, very much after the manner of our whinchat (*Motacilla rubetra*), but when alighting it frequently expands its tail like a fan. The sexes are exactly similar in size and plumage.

Mr. G. R. Gray observes, that the genus Muscisaxicola is probably synonymous with Lessonia of Mr. Swainson; but the latter name cannot be used, as it has already been twice employed in other branches of Natural History.

2. Muscisaxicola macloviana. *G. R. Gray.*

Sylvia macloviana, *Garn.* Voy. de la Coqu. Zool. p. 663.
Curruca macloviana, *Less.*

I brought home only one specimen of this bird; it came from East Falkland Island, whence also those described by Messrs. Lesson and Garnot were procured. Mr. Gould considered it a distinct species, but having carefully compared it with *M. mentalis*, I can see not the smallest difference in any point, excepting that it is somewhat larger in all its dimensions. The length of the whole body is .6 of an inch greater, of wing when folded .45, of tarsus .2, greater than in the foregoing species. I can scarcely hesitate in thinking it a large-sized local variety, from some favourable condition in the Falkland Islands to its growth.

3. Muscisaxicola brunnea. *Gould.*

M. griseo-fusca; gutture abdomineque albis flavescenti tinctis, pectore obscuro; alis caudâque obscure fuscis, singulis plumis rufescenti fusco marginatis; rectricum externarum radiis lateralibus flavescentibus.

Long. tot. 5 unc.; *rostri*, $\frac{1}{2}$; *alæ*, 3$\frac{1}{2}$; *caudæ*, 2$\frac{2}{3}$; *tarsi*, 1.

Head, and all the upper surface greyish brown; wings and tail dark brown, each feather margined with reddish brown; the outer webs of the external tail feathers buffy white; throat and all the under surface white, slightly tinged with buff; bill and feet blackish brown.

Habitat, Port St. Julian, Patagonia. (*January*).

The only specimen I procured was immature.

4. Muscisaxicola nigra. *G. R. Gray.*

Alauda nigra, *Boddater.*
——— rufa, *Gmel.*
——— fulva, *Lath.* Index.
Anthus fulvus, *Vieill.* Ency. Meth. p. 309.
——— variegatus, *Gerv. & Eydoux,* Mag. de Zool. 1836, p. 26.
Sylvia dorsalis, *King.*
Lessonia erythronotus, *Swains.* Class. of Birds.
Alouette noire à dos fauve, *Pl. enl.* 788.
L'Alouette à dos rouge, *Azara,* No. 149.

This bird has a wide geographical range. It is found in La Plata, Patagonia, Tierra del Fuego, and on the west coast at least as far north as the valley of Copiapó, in Northern Chile. It is every where common: it is a quiet, tame, inoffensive little bird: it lives on the ground, and frequents sand-dunes, beaches, and rocky coasts, which it seldom leaves: the broad shingly beds of the rivers in Chile have, however, tempted it inland, together with the Opetiorhynchus. I was told that it builds in low bushes.

1. ANTHUS CORRENDERA. *Vieill.*

Anthus correndera, *Vieill.* Ency. Meth. i. p. 325.
La correndera, *Azara*, No. 145.

This titlark is found in La Plata, Chile, and the Falkland Islands. I was informed by an intelligent sealer, that it is the only land-bird on Georgia and South Orkney (lat. 61° S.): it has, therefore, probably a further range southward than any other land-bird in the southern hemisphere. It does not live in flocks, is very common, and resembles a true Alauda in most of its habits. This species (as well as the following) is so closely allied to our meadow pipit, *Anthus pratensis*, that Latham considered it only as a variety; the latter has a high northern range, as the former has a southern one. There can be little doubt that the bird alluded to by Mr. Yarrell (British Birds, p. 392, vol. i.) as having been caught in the Southern Atlantic Ocean, nine hundred miles from Georgia, was this species, which was mistaken, owing to its close similarity, for the true *Anthus pratensis*.

2. ANTHUS FURCATUS. *D'Orb. & Lafr.*

A. furcatus, *D'Orb. & Lafr.* Mag. de Zool. 1836, p. 27. Voy. de l'Amer. Mer. Av. p. 227.

My specimens were procured on the northern bank of the Plata. It is more common there than the foregoing species, to which it is most closely allied: its chief distinguishing character appears to be the greater shortness of its toes and of the hind claw. I have seen this species alight on twigs. In the breeding season it flies upward, and then falls to the ground, with raised wings, in the peculiar manner common to the *Anthus arboreus* of England. It builds on the ground; nest simple; egg $\frac{10}{10}$ of an inch in length, and $\frac{7}{11}$ in width; colour dirty white, with small specks and blotches of dull red and obscurer ones of purple. This species, both in habits and structure, appears to be an analogue of *A. arboreus* of the northern hemisphere, as *A. correndera* is of *A. pratensis*. Mr. Yarrell informs me that the egg of *Anthus furcatus* is very different from that of *A. arboreus*, although the parent birds are so similar.

3. ANTHUS CHII. *Licht.*

A. Chii, *Licht. Spix.* Av. Sp. No. i. t. lxxvi. fig. 2. p. 75.
Le Chii, *Azara*, No. 146.

My specimen was procured at Rio de Janeiro, in Brazil.

Sylvicola aureola. *Gould.*

Plate XXVIII.

S. supra flavescenti olivacea; fronte cerviceque nitide flavis, singulis plumis ad apicem rufescenti castaneis; occipite griseo; alis caudâque nigrescentibus, latè flavo-marginatis; genis guttureque nitidè flavis; pectore concolori sed singulis plumis in medio pallidè castaneo notatis; abdomine albescenti.

Long. tot. 5 unc.; *rostri*, $\frac{6}{12}$; *alæ*, $2\frac{6}{12}$; *caudæ*, $2\frac{3}{12}$; *tarsi*, $\frac{10}{12}$.

The nape of the neck, back and tail-coverts yellowish olive; the wings and tail blackish, broadly margined with yellow; the front and crown yellow, with the tips of the feathers reddish castaneous; the hind head grey mixed with yellow, the cheeks and the throat bright yellow; the breast of the same colour, but each feather is marked down the middle with pale reddish castaneous, the sides and middle of the abdomen whitish.

Habitat, Galapagos Archipelago. (*September*).

This bird is not uncommon on these islands. It has the habits of our Sylviæ. It frequents the thickets in the lower, dry and rocky parts of the island, and especially a peculiar bush, with thick foliage, which grows only near the sea-coast.

Cyanotis omnicolor. *Swains.*

Regulus omnicolor, *Vieill.* Gal. pl. 166.
Sylvia rubrigastra, *Vieill.*
Regulus Byronensis, *Gray*, Griff. An. King. pl.
Tachuris omnicolor, *D'Orb. & Lafr.*
Tachuris roi, *Azara*, No. 161.

My specimens were obtained at Maldonado in June, and therefore probably it is not a bird of passage. It frequented reeds on the borders of a lake, but was exceedingly rare. I likewise saw one in Northern Patagonia, and in a collection of birds at Santiago, in Chile, made there by an inhabitant of the place. The soles of the feet of this exquisitely beautiful little bird are bright orange.

TRICHAS VELATA. *G. R. Gray.*

Sylvia velata, *Vieill.* Ois. de l'Amer. Sept. ii. pl. 74.
——— *D'Orb. & Lafr.* Mag. de Zool. 1836, p. 20.
Tanagra canicapilla, *Swains.* Ill. Orn. pl. 174.
Trichas canicapilla, *Swains.*

My specimen was procured at Maldonado in June.

FAMILY.—FRINGILLIDÆ.
SUB-FAM.—ALAUDINÆ.
MELANOCORYPHA CINCTURA. *Gould.*

M. Fæm. fuscescenti rufa; gulâ abdomineque medio pallidioribus; remigibus ad apicem nigrescenti fuscis; rectricibus singulis maculâ albâ ovatâ nigrescenti fuscâ ad apicem notatis.

Long. tot. 5¾ unc.; *alæ*, 3¼; *caudæ*, 2¼; *tarsi*, ⅔; *rost.* ½.

The whole of the plumage, bill, and feet, sandy rufous brown, which is lightest on the throat and centre of the abdomen; primaries near their extremities passing into blackish-brown; and each of the tail feathers with a large oval spot of blackish-brown near the tip.

Habitat, St. Jago, Cape Verde Islands. (*September.*)

This bird inhabits the most arid plains of lava; it runs, and in its habits resembles, in many respects, a lark.

PYRRHALAUDA NIGRICEPS. *Gould.*

P. supra fuscescenti alba, plumis medio obscurioribus; fronte, genis lineâque pectoris utrinque albis; corpore infra lineâque à basi rostri supra oculos ad occiput transiente nigris; caudæ plumis mediis nigrescentibus fuscescenti albo marginatis, plumis externis atris.

Long. tot. 4⅝ unc., *alæ*, 2⅞; *caudæ*, 1⅝; *tarsi*, $\frac{6}{16}$; *rost.* $\frac{4}{16}$.

Upper surface brownish-white, with the middle of the feathers darker; the front, cheeks, and a line on each side of the breast white; beneath the body, and a line from the bill passing over the eyes to the hind head, black; the tail with the middle feathers blackish, margined with brownish-white, the outer feathers deep black; the bill and feet pale.

Habitat, St. Jago, Cape Verde Islands. (*September and January.*)

Like the last species, this bird inhabits sterile lava plains; it runs like a lark, and generally goes in small flocks.

Sub-Fam.—PYRRHULINÆ.

Spermophila nigrogularis. *Gould.*

S. capite corporeque supra, alis caudáque fusco cinereis; loris guláque nigris; lineis à rostri angulis per collum utrinque descendentibus, pectore abdomineque mediis, tegminibusque caudalibus inferioribus cinereo albis.
Fem.? supra olivaceo fusca, subtus pallidior.

Long. tot. 3 unc.; *alæ*, $2\frac{1}{4}$; *caudæ*, 2; *tarsi*, $\frac{5}{8}$; *rostri*, $\frac{1}{10}$.

Male.—Head, all the upper surface, wings and tail, brownish-grey; lores and throat black; lines from the angle of the bill down each side of the neck, centre of the chest and abdomen, and the under tail coverts greyish-white; bill light horn colour; feet dark-brown.

Female?—The whole of the plumage olive-brown above, and lighter beneath; bill and feet brown.

Habitat, Monte Video. (*November.*)

1. Crithagra? brasiliensis.

Fringilla Brasiliensis, *Spix.* Av. Sp. Nov. ii. t. lxi. f. 1. m. 2. fem. p. 47.

My specimens were obtained from the northern bank of the Plata, in the months of June and November.

2. Crithagra? brevirostris. *Gould.*

C. vertice dorsoque pallidè olivaceo fuscis, plumis singulis striâ angustâ mediâ nigro-fuscâ, pennis scapularibus alis caudáque nigrofuscis cinereo olivaceo latè marginatis; uropygio virescenti flavo; loris, gulâ, pectore humero infra, abdomine, tegminibusque caudæ inferioribus latè flavis.

Long. tot. 5 unc.; *rostri*, $\frac{7}{16}$; *alæ*, $2\frac{7}{8}$; *caudæ*, $2\frac{1}{4}$; *tarsi*, $\frac{5}{8}$.

Crown of the head and back, light olive-brown, with a narrow stripe of blackish-brown in the centre of each feather; scapularies, wings and tail, blackish-brown, broadly margined with greyish olive; rump greenish-yellow; lores, throat, chest, under surface of the shoulders, abdomen, and under tail-coverts bright yellow : bill and feet brown.

Habitat, Maldonado (*May*), and Valparaiso (*September*).

Near Maldonado, I saw very large flocks of this species feeding on the open grassy plains. When the whole flock rises, these birds utter a low but shrill chirp. In Chile I obtained only one specimen.

Sub-Fam.—EMBERIZINÆ.

1. Emberiza gubernatrix. *Temm.*

Emberiza gubernatrix, *Temm.*, Pl. Col. 63 & 64.
——— cristata, *Swains*, Zool. Ill. pl. 148.
——— cristatella, *Vieill.* Gal. des Ois. pl. 67.
Yellow crested grosbeak, *Lath.* Hist.
La huppe jaune, *Azara*, No. 129.

My specimen was procured on the banks of the Parana, near Santa Fe, in latitude 31° S.

2. Emberiza luteoventris. *G. R. Gray.*

Fringilla luteoventris, *Meyen*, Nov. Act. 1880, pl. 12. f. 3.

This bird was procured at Santa Cruz, in Southern Patagonia; it was rare there.

Chrysometris campestris. *Gould.*

Fringilla campestris, *Spix.* Avium Nov. Sp. ii. p. 47, pl. 59. f. 3. ♀

C. Mas: olivaceus; dorsi plumis singulis flavo marginatis, uropygii præsertim; vertice, gulâ, alis caudâque nigris, alis caudâque plus minusve flavo-marginatis; capitis lateribus corporeque infra latè flavis.

Long. tot. 4 unc. 11 lin.; *rost.* 5 lin.; *alæ*, 2¾; *caudæ*, 2⅛; *tarsi*, 7 lin.

Male; olivaceous, with each feather of the back margined with yellow, especially on the rump; the top of the head, throat, wings and tail, black, the two latter margined more or less with yellow; the sides of the head and beneath the body bright yellow.

Habitat, forests of Tierra del Fuego (*February*), Valparaiso (*September*).

Sub-Fam.—FRINGILLINÆ.

1. Ammodramus longicaudatus. *Gould.*

Plate XXIX.

A. vertice humeroque cinereofuscis, dorso pallescenti fusco, uropygio rufescenti fusco tincto, plumis singulis strigâ mediâ fuscâ; tectricibus alarum majoribus, remigibus primariis secundariisque et caudâ nigrescentibus, cinereo albo externe marginatis; fronte, strigâ superciliari corporeque infra flavescentibus.

Long. tot. 5¾ unc.; *alœ*, 2⅜; *caudœ*, 3; *tarsi*, ¾; *rostri*, 9/16.

Crown of the head and shoulder, greyish brown; back, light brown, tinged with reddish brown on the rump, and with a stripe of dark brown down the centre of each feather; greater wing-coverts, primaries, secondaries, and tail blackish, margined externally with greyish white; forehead, stripe over the eye, and all the under surface, buff; bill black; feet brown. Young, or a bird after gaining its new plumage, differs in having the whole of the upper surface rich brown, with a tinge of olive and with a stripe of dark brown down each feather, and in having the wing coverts margined with reddish instead of greyish brown.

Habitat, Monte Video (*November*), Maldonado (*June*).

At Maldonado this bird frequented, in small flocks, reeds and other aquatic plants bordering lakes. In general habits, as well as in place of resort, it resembles those species of Synallaxis and Limnornis, with which it is often associated. It appears to live entirely on insects, and I found in the stomach of one which I opened various minute Coleoptera. Mr. Gould remarks, that the structure of this Ammodramus is very remarkable, for that it has a great general resemblance both in form and colouring to Synallaxis, although the thickness of its bill shows its relation to the Fringillinæ. In its habits it certainly is more allied to the former genus, than to its own family.

2. Ammodramus Manimbe, *G. R. Gray.*

Plate XXX.

Ammodramus xanthornus, in Plate, and in Gould's MS.
Fringilla Manimbè, *Licht.*, Cat. No. 253.
Emberiza Manimbè, *D'Orb. & Lafr.*, Syn. p. 77.
Manimbè, *Azara*, No. 141.

My specimen was obtained from Maldonado.

1. Zonotrichia matutina. *G. R. Gray.*

Fringilla matutina, *Licht.*, Cat. 25.
─────── *Kittl.* Kupfertafeln der Vögel, pl. 23. f. 3.
Tanagra ruficollis, *Spix*, Av. Sp. Nov. ii. t. liii. f. 3. p. 39.
Chingolo, *Azara*, No. 135. Chingolo Bunting, *Lath. Hist.*

I procured specimens of this species from the banks of the Plata, Bahia Blanca in Northern Patagonia, and from Valparaiso in Chile: in these countries it is perhaps the commonest bird. In the Cordillera, I have seen it at an elevation of at least 8000 feet. It generally prefers inhabited places, but it has not attained the air of domestication of the English sparrow, which bird in habits and general appearance it represents. It does not go in flocks, although several may be frequently seen feeding together. At Monte Video I found on the ground the nest of this species. It contained three eggs; these were .75 of an inch in length; form, rather rounded; colour, dirty white, with numerous small spots of chesnut and blackish brown, almost confluent towards the broadest end. It was in this nest that I found the parasitic egg, supposed to belong to a species of Molothrus, described in my journal.*

2. Zonotrichia canicapilla. *Gould.*

Z. vertice cinereo; loris regioneque paroticâ obscure fuscis: dorso collique lateribus rufis, dorso superiori et uropygio fuscis; dorso medio nigrescenti fusco, plumis singulis pallido fusco marginatis; tectricibus alarum nigrescenti fuscis, rufescente fusco marginatis, apice albis, duas fascias obliquas trans alarum formantibus.

Long. tot. 5¼ unc.; *alæ*, 2⅞; *caudæ*, 2½; *tarsi*, ⅞; *rostri*, ⅜.

Crown of the head grey; lores and ear-coverts dark brown; back and sides of the neck rufous; upper part of the back and rump brown; centre of the back blackish brown, each feather margined with light brown; wing-coverts blackish brown, margined with reddish brown, and tipped with white, forming two oblique bands across the wing; primaries, secondaries, and tail, dark brown, margined with greyish brown; throat and all the under surface brownish grey; and feet brown.

Habitat, Port Desire in Patagonia, and Tierra del Fuego.

This species is not uncommon in Tierra del Fuego, wherever there is any open

---

* Journal of Researches during the Voyage of the Beagle, p. 60.

space. Of the few birds inhabiting the desert plains of Patagonia, this is the most abundant. At Port Desire I found its nest : egg, about .83 in length ; form somewhat more elongated than in that of the last species ; colour, pale green, almost obscured by minute freckles and clouds of pale dull red.

3. ZONOTRICHIA STRIGICEPS. *Gould.*

Z. capite castaneo, lineâ mediâ obscure diviso, plumis singulis striâ mediâ nigrofuscâ, humeri flexurâ rufâ ; corpore supra fuscescente, plumis singulis striâ latâ mediâ obscure fuscâ ; remigibus, primariis caudâque nigro-fuscis pallidè fusco marginatis ; strigâ superciliari, faciei collique lateribus, gulâ pectore abdomineque medio cinereis ; hypochondriis tegminibusque caudæ inferioribus flavescentibus.

Long. tot. $5\frac{3}{4}$ unc. ; *rostri*, $\frac{1}{2}$; *alæ*, $2\frac{1}{2}$; *caudæ*, $2\frac{2}{3}$; *tarsi*, $\frac{3}{8}$.

Head chestnut, divided down the middle by a line of deep grey, each feather with a stripe of blackish brown down the centre ; point of the shoulder rufous ; the remainder of the upper surface light brown, with a broad stripe of dark brown down the centre of each feather ; primaries and tail brown ; secondaries blackish brown, margined all round with pale brown ; stripe over each eye, sides of the face and neck, throat, breast, and centre of the abdomen, grey ; flanks and under tail-coverts buff ; upper mandible black ; under mandible light horn colour ; feet brown.

Habitat, Santa Fe. Lat. 31° S. (*October.*)

This species appears to replace in this latitude the *Z. matutina*, which is so abundant on the banks of the Plata and in Chile, as that species does the *Z. canicapilla* of Patagonia and Tierra del Fuego.

PASSERINA JACARINA. *Vieill.*

Tanagra jacarina, *Linn.*
Passerina jacarina, *Vieill.* Ency. Meth. p. 933.
Emberiza jacarina, *D'Orb. & Lafr.*, Syn.
Le Sauteur, *Azara*, No. 138.
Euphone jacarina, *Licht.* Cat. p. 30.
Fringilla splendens, *Vieill.* Ency. p. 981 ?

I procured a specimen of this bird at Rio de Janeiro.

1. Fringilla Diuca. *Mol.*

Fringilla Diuca, *Kittl.* Mem. de St. Petersb. t. i. pl. 11.
———————— Mag. de Zool. 1837, pl. 69.
Emberiza Diuca, *D'Orb. et Lafr.* Syn. Mag. of Zool. 1838, f. 77.

This bird is very common on the coast of Chile, from the humid forests of Chiloe to the desert mountains of Copiapó. In Chiloe it is perhaps the most abundant of the land birds; south of Chiloe I never saw it, although the nature of the country does not change them. On the eastern side of the continent, I met with this bird only at the Rio Negro, in northern Patagonia. I do not believe it inhabits the shores of the Plata, although so common in the open country, under corresponding latitudes west of the Cordillera. The Diuca, as this Fringilla is called in Chile, generally moves in small flocks, and frequents, although not exclusively, cultivated ground in the neighbourhood of houses: habits very similar to those of the *Zonotrichia matutina*. During incubation, the male utters two or three pleasing notes, which Molina has in an exaggerated description called a fine song. In October, at Valparaiso, I found the nest of this bird in the trellis-work of a vineyard, close by a much frequented path. The nest is shallow, and about six inches across; the outer part is very coarse, and composed of the thin stalks of twining plants, strengthened by the husky calices of a composite flower; this outside part is lined by many pieces of rag, thread, string, tow, and a few feathers. Eggs rather pointed, oval, ·94 of an inch in length; colour, pale dirty green, thickly blotched by rather pale dull-brown, which small blotches and spots become confluent, and entirely colour the broad end.

2. Fringilla Gayi. *Eyd. & Gerv.*

Fringilla Gayi, *Eyd. & Gerv.* Mag. de Zool. 1834. pl. 23.
Emberiza Gayi, var. *D'Orb. & Lafr.* Syn. p. 76.

This Fringilla, which was first brought from Chile, is abundant in the southern parts of Patagonia.

3. Fringilla formosa. *Gould.*

F. fronte lorisque nigris; vertice, genis, gulâ, alarum tegminibus cæruleo griseis, tegminibus primariis, secundariis rectricibusque griseo-nigris, cærulescenti-griseo marginatis, dorso flavescenti castaneo; tegminibus caudalibus inferioribus pallidè griseis; uropygio pectore abdomine hypochondriisque saturatè flavis.

Long. tot. 5½ unc.; *alæ*, 3¼; *caudæ*, 2⅜; *tarsi*, ¾; *rostri*, ⅓.

Forehead and lores black; crown of the head, sides of the face, throat, wing coverts, and the margins of the primaries, secondaries, and tail feathers, blue grey; the remainder of the primaries, secondaries, and tail feathers, greyish-black; back yellowish-chestnut; under tail coverts light-grey; rump, breast, abdomen and flanks, deep wax-yellow; bill bluish horn-colour; feet light brown.

Habitat, Tierra del Fuego (*December and February*).

This finch is common on the outskirts of the forests in Tierra del Fuego. Mr. Gould remarks, that it is nearly allied to *F. Gayi*, but it is much smaller, and is richer in its colouring.

4. FRINGILLA FRUTICETI, *Kittl.*

Fringilla fruticeti, *Kittl.* Kupf. der Vögel, pl. 23. f. 1.
Emberiza luctuosa, *Eyd. et Gerv.* Mag. de Zool. 1834. Cl. 11. pl. 71.
———— *D'Orb. et Lafr.* Syn. p. 80.

I obtained specimens of this bird from Northern Chile, and Southern Patagonia. I saw it also in the Cordillera of Central Chile, at an elevation of at least eight thousand feet, near the upper limit of vegetation. In Patagonia it is not common, it frequents bushy valleys in small flocks, from six to ten in number. These birds sometimes move from thicket to thicket with a peculiar soaring flight: they occasionally utter very singular and pleasing notes.

5. FRINGILLA CARBONARIA. *G. R. Gray.*

Emberiza carbonaria, *D'Orb. et Lafr.* Synop. p. 79.

I never saw this bird but once, and then it was in small flocks, on the most desert parts of the plains between the rivers Negro and Colorado, in Northern Patagonia.

6. FRINGILLA ALAUDINA. *Kittl.*

Fringilla alaudina, Kupf. der Vögel, pl. 23. f. 2.
Emberiza guttata, *Meyen,* Nov. Act. Cur. xvii. pl. 12.
———— *D'Orb. & Lafr.* Syn. p. 78: Adult.
Passerina guttata, *Eyd. & Gerv.* Mag. de Zool. 1834. pl. 70. p. 22.

My specimens were obtained from the neighbourhood of Valparaiso.

1. Passer Jagoensis. *Gould.*

PLATE XXXI.

Pyrgita Jagoensis, *Gould*, Proc. of Zool. Soc. 1837. p. 77.

P. summo capite, et maculâ parvâ gulari intensè nigrescenti-fuscis; strigâ superciliari, collo, humeris dorsoque intensè castaneis, hujus plumis strigâ fuscâ centrali notatis; alis caudâque brunneis, tectricibus alarum minoribus albis, qui color fasciam transversam efficit: lineâ angustâ albâ à nare ad oculum; genis corporeque subtùs albis, hoc colore in cinereum ad latera transeunte: rostro, pedibusque fuscis.

Long. tot. 5 unc.; *caudæ*, 2¼; *alæ*, 2¼; *rost*. ½; *tarsi*, ¾.

Crown of the head and a small mark on the throat intense blackish brown, with a stripe on the eyebrows, the neck, shoulders and back bright chestnut, the feathers of the latter marked with a central dusky streak; wings and tail brown, with the smaller wing coverts white, forming a transverse bar; a narrow white line from the nostrils to the eye; cheeks and under side of body white, this colour passing into grey on the sides; beak and feet dusky.

Habitat, St. Jago, Cape Verde Islands (*January*).

This is the commonest bird in the island; it frequents, generally in small flocks, both the neighbourhood of houses and wild uninhabited spots. It was building its nest towards the end of August.

2. Passer Hispaniolensis. *G. R. Gray.*

Fringilla Hispaniolensis, *Temm. Man.* i. 353.

In the month of January I obtained a specimen of this bird from St. Jago, one of the Cape Verde Islands, where it was not common.

1. Chlorospiza? melanodera. *G. R. Gray.*

PLATE XXXII.

Emberiza melanodera, *Quoy & Gaim.* Voy. de L'Uranie, Zool. i. p. 109.

C. flavescenti olivacea; dorso superiori cinereo rufoque mixto; vertice, auribus, colli lateribus pectoreque cinereis rufomixtis; lineâ à naribus pone oculos transiente genisque albis; plumis inter rostrum et oculos gulâque atris; remigibus primariis et secundariis nigrescentibus flavo marginatis: caudâ rectricibus mediis olivaceo-fuscis, tribus externis ferè toto pallidè flavis; abdomine medio flavescenti albo, lateribus obscurioribus.

Long. tot. 6½ unc.; *alæ*, 3¼; *caudæ*, 2¾; *tarsi*, 10 lines; *rostri*, 5 lin.

Adult. Yellowish olive, mixed with grey and rufous on the upper part of the back;

top of the head, ears, sides of the neck and breast, grey mixed with rufous; the lines from the nostrils reaching behind the eyes and cheeks, white; the space between the bill and eye, and the throat, deep black; the primaries and secondaries blackish, margined with yellow; the tail, with the middle feathers, olivaceous black, with the three external nearly wholly pale yellow; the middle of the abdomen yellowish white, with the flanks darker.

Young: Upper surface brownish white, with the middle of each feather black; the throat lighter; the wing coverts and secondaries margined with white and brown; the primaries with yellow; the tail blackish, with their outer margins yellow, and the external feather wholly pale yellow white; beneath the body pale yellowish white, streaked on the breast and flanks with a darker tint.

Habitat, East Falkland Island (*March*), and Santa Cruz, Patagonia (*April*).

This bird is extremely abundant in large scattered flocks in the Falkland Islands.

2. CHLOROSPIZA? XANTHOGRAMMA. *G. R. Gray.*

PLATE XXXIII.

C. cinerascenti olivacea, rufo paulo tincta; lineâ à naribus pone oculos transiente genisque flavis; plumis inter rostrum et oculos gulâque atris; remigibus secundariis nigrescentibus, cinereo et olivaceo latè marginatis; primariis nigrescentibus, flavo angustè marginatis; caudâ cinerascenti nigrâ, plumis externis albis; corpore infra flavescenti albo, hypochondriis obscurioribus.

Long. tot. 7⅝ unc.; *alæ*, 3⅞; *caudæ*, 3; *tarsi*, 1; *rostri*, 7 lin.

Adult: Greyish olive, very slightly mixed with rufous, a line from the nostrils reaching behind the eyes and cheeks, yellow; the space between the bill and eye, and the entire throat, deep black; the secondaries blackish, broadly margined with grey and olive; the primaries blackish, slightly margined with yellow; the tail greyish black, with the outer feathers white; beneath the body yellowish white, darker on the flanks.

Female: Upper surface brownish white, with each feather blackish brown in the middle, the head and throat paler; the wing-coverts and secondaries blackish, margined with brownish white; the primaries blackish, slightly margined with yellow; the tail blackish white-margined, with the outer feathers nearly wholly white; beneath the body yellowish white, streaked with brown on the breast and flanks: and the space from the nostrils reaching to behind the eyes and cheeks, yellowish.

Habitat, East Falkland Island (*March*), and Tierra del Fuego (*February*).

This species is common at the Falkland Islands, and it often occurs mingled in the same flock with the last one. I suspect, however, it more commonly frequents higher parts of the hills. These species have a very close general resemblance; but the marks about the head, which are white in the *C. melanodera*, are yellow in the *C. xanthogramma*, while the parts of the tail-feathers which are white in the latter, are yellow in the *C. melanodera:* this difference of colours does not hold in the females, but they may be at once distinguished by the greater length of wing, when folded, of the *C. xanthogramma*.

CHRYSOMITRIS MAGELLANICA. *Bonap.*

Fringilla Magellanica, *Vieill.* Ency. Meth. 983; Ois. Chant. de la Zône Torride, pl. 30;
    Audubon, Birds of Am. pl. 394, f. 2.
Gafarron, *Azara*, No. 134.
Fringilla icterica, *Licht.* Cat. p. 26.

This bird was very abundant in large flocks during May, at Maldonado; I found it also at the Rio Negro.

SUB-FAMILY.—TANAGRINÆ.

PITYLUS SUPERCILIARIS.

Tanagra superciliaris, *Spix.* Av. Sp. Nov. 2. t. lvii. fig. 1. p. 44.

My specimen was procured from Santa Fé, in Lat. 31° S.

1. AGLAIA STRIATA. *D'Orb. & Lafr.*
PLATE XXXIV.

♂ Tanagra striata, *Gmel.* Syst. 1. 899; Ency. Meth. 776; *Licht.* Cat. p. 31. Sp. 347;
    Proc. Zool. Soc. 1837, p. 121, pl. 34 of this work.
L'Onglet, *Buff.* iv. p. 256.
Le Lindobleu, dore et noir, *Azara*, No. 94.
♀ Tanagra Darwinii, *Bonap.*; Proc. Zool. Soc. 1837, p. 121.

I saw the only specimen, which I procured, feeding on the fruit of an opuntia at Maldonado.

Mr. G. R. Gray is induced to consider the species figured under the name of *T. Darwinii*, as the *T. striata*, *Gm.* and the *T. Darwinii* of the Zoological Society's Proceedings, as the female of the same species, while the young birds may be described as following:

Brown, with the margins of the dorsal feathers greenish-brown, those of the
    wings and tail margined brownish-white; head and neck greyish-green;

beneath the body pale dusky green, somewhat darker on the breast and sides; uropygium yellowish-green.

Three specimens of this species are contained in the British Museum, exhibiting male, female, and young.

2. AGLAIA VITTATA.

Tanagra vittata, *Temm.* Pl. col.

Maldonado; not common.

PIPILLO PERSONATA. *Swains.*
PLATE XXXV.

P. personata, *Swains.* Two cent. and a quart. p. 311.

Maldonado; not common. The stomach of one, which I shot, contained seeds.

EMBERIZOIDES POLIOCEPHALUS. *G. R. Gray.*

E. olivaceus, dorsi plumis medio nigro striatis; capite guláque cinereis, priore plumis singulis, medio nigrostriatis; corpore infra rufescenti albo; hypochondriis tectricibus caudæ inferioribus obscurioribus; alarum margine latè flava, remigibus primariis secundariisque nigris; prioribus pallidè olivaceo, posterioribus olivascente flavo latè marginatis.

Long. tot. 7¾ unc.; *alæ,* 3½; *caudæ,* 4; *tarsi,* 1¼; *rostri,* 8 lin.

Olivaceous, with the feathers of the back marked down the middle with black; the head and throat cinereous, with each feather of the former streaked down the middle with black; beneath the body rufous white, darker on the flanks and under tail coverts; the border of the wings bright yellow; the secondaries and primaries black, the former broadly margined with pale olive, the latter with bright olivaceous yellow; base of bill dusky orange.

Habitat, northern shore of the Plata. (*May* and *August.*)

This bird is common both near Monte Video and Maldonado, in swamps. Stomach full of seeds: it makes a shrill loud cry: its flight is clumsy, as if its tail were disjointed.

FAM.—COCCOTHRAUSTINÆ.
GENUS, GEOSPIZA, *Gould.*

Corporis figura brevissima et robusta.

Rostrum magnum, robustum, validum, altitudine longitudinem præstante; culmine arcuato et capitis verticem superante, apice sine denticulo, lateribus tumidis.

Naribus basalibus et semitectis plumis frontalibus.

Mandibulâ superiori tomiis medium versus sinum exhibentibus, ad mandibulæ inferioris processum recipiendum. Mandibula inferior ad basin lata, hoc infra oculos tendente. Alæ mediocres remige primo paulo breviore secundo, hoc longissimo.
Cauda brevissima et æqualis.
Tarsi magni et validi, digito postico, cum ungue robusto et digito intermedio breviore; digitis externis inter se æqualibus at digito postico brevioribus. Color in maribus niger, in fæm. fuscus.

This singular genus[*] appears to be confined to the islands of the Galapagos Archipelago. It is very numerous, both in individuals and in species, so that it forms the most striking feature in their ornithology. The characters of the species of Geospiza, as well as of the following allied subgenera, run closely into each other in a most remarkable manner.

In my Journal of Researches, p. 475, I have given my reasons for believing that in some cases the separate islands possess their own representatives of the different species, and this almost necessarily would cause a fine gradation in their characters. Unfortunately I did not suspect this fact until it was too late to distinguish the specimens from the different islands of the group; but from the collection made for Captain FitzRoy, I have been able in some small measure to rectify this omission.

In each species of these genera a perfect gradation in colouring might, I think, be formed from one jet black to another pale brown. My observations showed that the former were invariably the males; but Mr. Bynoe, the surgeon of the Beagle, who opened many specimens, assured me that he found two quite black specimens of one of the smaller species of Geospiza, which certainly were females: this, however, undoubtedly is an exception to the general fact; and is analogous to those cases, which Mr. Blyth[*] has recorded of female linnets and some other birds, in a state of high constitutional vigour, assuming the brighter plumage of the male. The jet black birds, in cases where there could be no doubt in regard to the species, were in singularly few proportional numbers to the brown ones: I can only account for this by the supposition that the intense black colour is attained only by three-year-old birds. I may here mention, that the time of year (beginning of October) in which my collection was made, probably corresponds, as far as the purposes of incubation are concerned, with our autumn. The several species of Geospiza are undistinguishable from each other in habits; they often form, together with the species of the following subgenera, and likewise with doves, large irregular flocks. They frequent the rocky and extremely arid parts of the land sparingly covered with almost naked bushes, near the coasts;

---

[*] This genus, and the following sub-genera, were named by Mr. Gould at a meeting of the Zool. Soc. Jan. 10 1837, p. 4. of Proceedings.

† Remarks on the Plumage of Birds, Charlesworth's Mag. of Nat. History, vol. i. p. 480.

for here they find, by scratching in the cindery soil with their powerful beaks and claws, the seeds of grasses and other plants, which rapidly spring up during the short rainy season, and as rapidly disappear. They often eat small portions of the succulent leaves of the *Opuntia Galapageia*, probably for the sake of the moisture contained in them : in this dry climate the birds suffer much from the want of water, and these finches, as well as others, daily crowd round the small and scanty wells, which are found on some of the islands. I seldom, however, saw these birds in the upper and damp region, which supports a thriving vegetation; excepting on the cleared and cultivated fields near the houses in Charles Island, where, as I was informed by the colonists, they do much injury by digging up roots and seeds from a depth of even six inches.

1. Geospiza magnirostris. *Gould.*
Plate XXXVI.

G. fuliginosa, crisso cinerascenti-albo; rostro nigro-brunnescente lavato; pedibus nigris.

Long. tot. 6 unc. ; alæ, 3½ : caudæ, 2 ; tarsi, 1 ; rostri, $\frac{7}{8}$; alt. rost. 1.

Fœm. vel Mas jun.; *corpore intensè fusco singulis plumis olivaceo cinctis; abdomine pallidiore ; crisso cinerascenti-albo ; pedibus et rostro, ut in mare adulto.*

Sooty black ; with the vent cinereous white, the bill black, washed with brownish, and the feet black.

Female, or young male : Deep fuscous, with each feather margined with olive, the abdomen much paler, with the under tail-coverts cinereous white, the feet and bill like those of the male.

Habitat, Galapagos Archipelago. (Charles and Chatham Islands.)

I have strong reasons for believing this species is not found in James's Island. Mr. Gould considers the *G. magnirostris* as the type of the genus.

2. Geospiza strenua. *Gould.*
Plate XXXVII.

G. fuliginosa, crisso albo, rostro fusco et nigro tincto ; pedibus nigris.

Long. tot. 5½ unc. ; alæ, 3 ; caudæ, 1$\frac{3}{8}$; tarsi, $\frac{3}{4}$; rostri, $\frac{3}{8}$; alt. rost. $\frac{3}{8}$.

Fœm. *Summo corpore fusco singulis plumis alarum caudæque plumis exceptis, pallidè cinerascenti-olivaceo cinctis ; gulâ et pectore fuscis ; abdomine lateribus et crisso pallidè cinerascenti-fuscis; rostro brunnescente.*

Sooty black, with the under tail coverts white ; the bill brown, tinged with black, and the feet black.

Female : Upper part of the body fuscous, with the margins of each feather, except those of the wings and tail, pale cinereous-olive; the throat and breast fuscous: the abdomen, sides, and under tail-coverts pale cinereous-fuscous; the bill brownish.

Habitat, Galapagos Archipelago (James and Chatham Islands.)

Geospiza fortis. *Gould.*
Plate XXVIII.

G. intense fuliginosa, crisso albo; rostro rufescenti-brunneo, tincto nigro; pedibus nigris.

Fœm. (vel Mas jun.) *Corpore suprà pectore et gutture intensè fuscis, singulis plumis cinerascenti-olivaceo marginatis; abdomine crissoque pallidè cinerascenti-brunneis; rostro rufescenti-fusco ad apicem flavescente; pedibus ut in mare.*

Long. tot. 4¾ unc.; *alæ*, 3; *caudæ*, 1½; *tarsi*, 1⅔; *rostri*, 7⁄12.

Deep sooty black; with the under tail-coverts and the bill reddish brown tinged with black; the feet black.

Female (or young male): The body above, breast and throat, deep fuscous, with each feather margined with cinereous-olive: the abdomen, and under tail-coverts pale cinereous-brown; the bill reddish fuscous, with the apex yellowish, and the feet like those in the male.

Habitat, Galapagos Archipelago, (Charles and Chatham Islands.)

4. Geospiza nebulosa. *Gould.*

G. summo capite et corpore nigrescenti-fuscis; singulis plumis cinerascenti-olivaceo marginatis; corpore subtus pallidiore, abdomine imo crissoque cinerascentibus; rostro et pedibus intensè fuscis.

Long. tot. 5 unc.; *alæ*, 2¾; *caudæ*, 1¾; *tarsi*, ¾; *rostri*, ⅝; alt. rost. ⅓.

Male.—Upper part of the head and body blackish fuscous, with each feather margined with cinereous olive; the body beneath paler, with the lowest part of the abdomen and under tail-coverts ashy; the bill and feet deep fuscous.

Habitat, Galapagos Archipelago, (Charles Island.)

5. Geospiza fuliginosa. *Gould.*

G. intensè fuliginosa, crisso albo, rostro fusco; pedibus nigrescenti-fuscis.

Long. tot. 4½ unc.; *alæ*, 2½; *caudæ*, 1⅝; *tarsi*, ¾; *rostri*, 1½; alt. *rostri*, ⅔.

Fœm. *Summo corpore, alis, caudáque intensè fuscis; singulis plumis cinerascenti-ferrugineo marginatis; corpore infra cinereo, singulis plumis medium versus obscurioribus; rostro brunneo; pedibus nigrescenti-brunneis.*

Deep sooty black, with the under tail coverts white; the bill fuscous, and the feet blackish fuscous.

Female: Upper part of the body; the wings and tail deep fuscous, with each feather margined with ashy ferrugineous; beneath the body cinereous, with each feather towards the middle darker; the bill brown, and the feet blackish brown.

Habitat, Galapagos Archipelago. (Chatham and James' Island.)

6. GEOSPIZA DENTIROSTRIS. *Gould.*

G. (Fœm. vel Mas jun.) *mandibulæ superioris margine in dentem producto, vertice corporeque supra fuscis; singulis plumis medium versus obscurioribus; secundariis tectricibusque alarum ad marginem stramineis; gutture et pectore pallidè brunneis, singulis plumis medium versus obscurioribus, imo abdomine crissoque cinerascenti-albis; rostro rufo-fusco; pedibus obscurè plumbeis.*

Long. tot. 4¾ unc.; *alæ*, 2⅔; *caudæ*, 1¾; *rostri*, ½; alt. rost. ⅜.

The margin of the upper mandible produced into a tooth; the vertex and above the body fuscous, with each feather towards the middle darker; the margins of the secondaries and wing coverts straw colour; the throat and breast pale brown, darker towards the middle of each feather; the sides and under tail-coverts cinereous white; the bill rufous fuscous, and the feet obscure lead colour.

Habitat, Galapagos Archipelago.

Mr. Gould considered this specimen a female, from the appearance of its plumage; but from dissection, I thought it was a male.

7. GEOSPIZA PARVULA. *Gould.*
PLATE XXXIX.

G. (Mas) *capite, gutture, et dorso fuliginosis; uropygio cinerascenti-olivaceo; caudâ et alis nigrescenti brunneis; singulis plumis caudæ et alarum, cinereo-marginatis; lateribus olivaceis, fusco guttatis; abdomine et crisso albis, rostro et pedibus nigrescenti-brunneis.*

Long. tot. 4 unc.; *alæ*, 2⅔; *caudæ*, 1½; *tarsi*, ¾; *rostri*, ⅜; alt. rost. ₅⁄₁₆.

Fœm. *Summo capite et dorso cinerascenti-brunneis, gutture, pectore, abdomine crissoque pallidè cinereis, stramineo tinctis.*

The head, throat, and back, sooty black; the lower part of the back cinereous olive; the tail and wings blackish brown, margined with cinereous; the sides olive with fuscous spots; the abdomen and under tail-coverts white; the bill and feet blackish brown.

Female: The upper surface cinereous brown; the throat, breast, abdomen, and the under tail coverts, pale cinereous tinged with straw colour.

Habitat, Galapagos Archipelago. (James' Island.)

BIRDS. 103

8. Geospiza dubia. *Gould.*

G. (Fœm. Mas ignot.) *summo capite et corpore suprà fuscis, singulis plumis cinerascenti-olivaceo marginatis; strigâ superciliari, genis, gutture, corpore infrà cinerascenti-olivaceis, singulis plumis notâ centrali fuscâ; alis caudâque brunneis singulis plumis olivaceo-cinereo marginatis; rostro sordidè albo, pedibus obscurè fuscis.*

Long. tot. $3\frac{3}{8}$ unc.; *alæ*, $2\frac{3}{8}$; *caudæ*, $1\frac{3}{8}$; *tarsi*, $\frac{7}{8}$; *rostri*, $\frac{5}{8}$: alt. *rostri*, $\frac{3}{8}$.

Upper surface fuscous, with each feather margined with cinereous olive; the streak above the eye, cheeks, throat, and beneath the body, cinereous olive, with the middle of each feather fuscous; the wings and tail brown, with each feather margined with cinereous ash; the bill white, and the feet obscure fuscous.

Habitat, Galapagos Archipelago, (Chatham Island.)

Sub-Genus.—CAMARHYNCHUS. *Gould.*

Camarhynchus *differt a genere* Geospiza, *rostro debiliore, margine mandibulæ superioris minùs indentato; culmine minùs elevato in frontem et plus arcuato; lateribus tumidioribus; mandibulâ inferiore minus in genas tendente.*

Camarhynchus psittaculus is the typical species.

1. Camarhynchus psittaculus. *Gould.*

Plate XL.

C. (Fœm.) *summo capite corporeque superiore fuscis; alis caudâque obscurioribus; gutture corporeque inferiore, cinerascenti-albis, stramineo tinctis; rostro pallidè flavescenti-fusco; pedibus fuscis.*

Long. tot. $4\frac{3}{4}$ unc.; *alæ*, $2\frac{3}{4}$: *caudæ*, $1\frac{3}{4}$; *tarsi*, $\frac{7}{8}$; *rostri*, $\frac{1}{2}$; alt. *rostri*, $\frac{1}{2}$.

The upper part of the head and body fuscous; the wing and tail darker; the throat, and beneath the body cinereous white, tinged with straw-colour; the bill pale yellowish fuscous, and the feet fuscous.

Habitat, Galapagos Archipelago, (James' Island.)

The species of Camarhynchus do not differ in habits from those of Geospiza; and the *C. psittaculus* might often be seen mingled in considerable numbers in the same flock with the latter. Mr. Bynoe procured a blackish specimen, which, doubtless, was an old male; I saw several somewhat dusky, especially about the head.

2. Camarhynchus crassirostris. *Gould.*

Plate XLI.

C. (Mas jun. et Fœm.) *corpore superiore intensè brunneo, singulis plumis cinerascenti-*

olivaceo marginatis; gutture pectoreque cinerascenti-olivaceis, singulis in medio plumis obscurioribus; abdomine, lateribus crissoque cinereis stramineo tinctis.

Long. tot. 5½ unc.; *alæ*, 3¾; *caudæ*, 2; *tarsi*, 1⅛; *rostri*, ½; alt. *rostri*, ½.

Upper part of the body deep brown, with each feather margined with cinereous olive; the throat and breast cinereous olive, with the middle of each feather darker; the abdomen, sides, and under tail coverts cinereous tinged with straw colour.

Habitat, Galapagos Archipelago, (Charles Island?)

I am nearly certain that this species is not found in James Island. I believe it came from Charles Island, and probably there replaces the *C. psittaculus* of James Island. I obtained three specimens, one male, and two females; from the analogy of so many species in this group, I do not doubt the old male would be black.

Sub-Genus.—CACTORNIS. *Gould*.

CACTORNIS *differt a genere* GEOSPIZA *rostro elongato, acuto, compresso, longitudine altitudinem excellente; mundibulæ superioris margine vix indentato; naribus basalibus et vix tectis; tarsis brevioribus, unguibus majoribus et plus curvatis.*

Cactornis scandens is the typical species.

1. CACTORNIS SCANDENS. *Gould*.
Plate XLII.

C. intensè fuliginosa, crisso albo; rostro et pedibus nigrescenti-brunneis.

Long. tot. 5 unc.; *rostri*, ¾; *alæ*, 2⅘; *caudæ*, 1¾; *tarsi*, ¾.

Fœm. *Corpore superiore, gutture pectoreque intensè brunneis, singulis plumis pallidiorè marginatis; abdomine crissoque cinereis, stramineo tinctis; rostro pallidè fusco; pedibus nigrescenti-fuscis.*

Deep sooty black, with the under tail-coverts white; the bill and feet blackish-brown.

Female: Upper surface of the body, throat and breast intensely brown, with the margins of each feather paler; the abdomen and the under tail coverts cinereous, tinged with straw-colour; the bill pale fuscous, and the feet blackish fuscous.

Habitat, Galapagos Archipelago, (James' Island.)

The species of this sub-genus alone can be distinguished in habits from the several foregoing ones belonging to Geospiza and Camarhynchus. Their most frequent resort is the *Opuntia Galapageia*, about the fleshy leaves of which they hop and climb, even with their back downwards, whilst feeding with their sharp beaks, both on the fruit and flowers. Often, however, they alight on the ground, and mingled with the flock of the above mentioned species, they search for seeds in the parched volcanic soil. The extreme scarceness of the jet-black specimens, which I mentioned under the head of the genus *Geospiza*, is well exemplified in the case of the *C. scandens*, for although I daily saw many brown-coloured ones, (and two collectors were looking out for them), only one, besides that which is figured, was procured, and I did not see a second.

2. CACTORNIS ASSIMILIS. *Gould*.

PLATE XLIII.

TISSERIN DES GALLAPAGOS, (île St. Charles,) *Neboux*, Revue Zoologique, 1840, p. 291.

C. Mas (jun?) *corpore suprà fuliginoso, (gutture abdomineque exceptis,) cinereo marginatis; rostro pallidè rufescenti-brunneo; pedibus nigrescenti-brunneis.*

Long. tot. 5¼ unc.; *rostri*, ⅜; *alæ*, 2¾; *caudæ*, 1¾; *tarsi*, ¾.

Upper surface of the body sooty black, margined with cinereous, as well as the throat and abdomen; the bill pale rufous brown; the feet blackish brown.

Habitat, Galapagos Archipelago.

I do not know from which island of the group this species was procured; almost certainly not from James Island. Analogy would in this case, as in that of *Camarhynchus crassirostris*, lead to the belief that the old male would be jet black. By a mistake this bird has been figured standing on the *Opuntia Darwinii*, a plant from Patagonia, instead of the *O. Galapageia*. I may here mention that a third and well characterized species of Cactornis has lately been sent by Captain Belcher, R.N. to the Zoological Society; as Capt. Belcher visited Cocos Island, which is the nearest land to the Galapagos Archipelago, being less than 400 miles distant, it is very probable that the species came thence.

SUB-GENUS.—CERTHIDEA. *Gould*.

CERTHIDEA *differt a genere* GEOSPIZA *rostro graciliore et acutiore; naribus basalibus et non tectis; mandibulæ superioris margine recto; tarsis longioribus et gracilioribus.*

Of the foregoing sub-genera, Geospiza, Camarhynchus and Cactornis belong to one type, but with regard to Certhidea, although Mr. Gould confidently believes it should also be referred to the same division, yet as in its slighter form and weaker bill, it has so much the appearance of a member of the *Sylviadæ*, he would by no means insist upon the above view being adopted, until the matter shall have been more fully investigated.

CERTHIDEA OLIVACEA. *Gould.*
PLATE XLIV.

C. summo capite, corpore superiore, alis caudâque olivaceo-brunneis; gutture et corpore infra cinereis; rostro pedibusque pallidè brunneis.

Long. tot. 4 unc.; *rostri*, ½; *alæ*, 2; *caudæ*, 1½; *tarsi*, ¾.

Upper part of the head, body, wings and tail, olivaceous brown; the throat, and beneath the body, cinereous; the bill and feet pale brown.

Habitat, Galapagos Archipelago. (Chatham and James Island).

I believe my specimens, which include both sexes, were procured from Chatham and James Islands; it is certainly found at the latter.

PHYTOTOMA RARA. *Mol.*

P. Bloxami, *Children*, Jard. and Selby's Ill.
P. rutila, *Vieill.* Mag. de Zool. 1832, ii. pl. 5.
P. silens, *Kittl.* Mem. de l'Acad. des Sci. de St. Petersb.

This is not a very uncommon bird in Central Chile: the farmers complain that it is very destructive to the buds of fruit trees. It is quiet and solitary, and haunts hedge-rows or bushes; its manners are similar to those of our bullfinch, (*Loxia Pyrrhula*). Iris bright scarlet. Mr. Eyton has given an anatomical description of this bird in the Appendix.

DOLICHONYX ORYZIVORUS. *Swains.*

Dolichonyx oryzivorus, *Swains.* Faun. Bor. Am. 2. 278.
Emberiza oryzivorus, *Linn.*

This one specimen only was seen at James Island, in the Galapagos Archipelago, during the beginning of October. It is remarkable that a bird migrating, according to Richardson, as far as 54° N. in North America, and generally inhabiting marshy grounds, should be found on these dry rocky islands under the equator. Mr. Gray and myself carefully compared this specimen with one from North America, and we could not perceive the slightest difference.

1. XANTHORNUS CHRYSOPTERUS. *G. R. Gray.*

Oriolus cayennensis, *Linn.* Syst. 1. 168 ?
Agelaius chrysopterus, *Vieill.*
Psarocolius chrysopterus, *Wagl* Syst. Av. p.

This bird generally frequents marshy grounds. I procured specimens from La Plata and from Chile; in the latter country it extends at least as far north as the valley of Copiapo, in 27° 20′: on the eastern plains it does not range, according to Azara, north of 28°. It builds in reeds. Molina says it is called by the Indians Thili, or Chile—hence he derives the name of the country.

2. XANTHORNUS FLAVUS. *G. R. Gray.*
PLATE XLV.

Oriolus flavus, *Gmel.*
Psarocolius flaviceps, *Wagl.* Syst. Avium.
Troupiale à tête jaune, *Azara*, No. 66.

This species is common at Maldonado in large flocks.

LEISTES ANTICUS. *G. R. Gray.*

Icterus anticus, *Licht.* Cat. p. 19.
Agelaius virescens, *Vieill.* Ency. Meth. 543.
Psarocolius anticus, *Wagl.*
Le Dragon, *Azara*, No. 65.

This bird is exceedingly abundant in large flocks on the grassy plains of La Plata. It is noisy, and in its habits resembles our starling.

1. AGELAIUS FRINGILLARIUS. *G. R. Gray.*

Icterus fringillarius, *Spix*, Av. Sp. No. 1. t. lxv. fig. 1 & 2. p. 68.
Psarocolius sericeus, juv., *Wagl.*

This species is rare at Maldonado, but appears more common on the banks of Parana in Lat. 31°. S. Spix says (vol. i. p. 68, Birds of Brazil), it is found in Minas Geraes.

2. AGELAIUS CHOPI. *Vieill.*

Turdus curæus, *Gmel.*
Le Chopi, *Azara*, No. 62.
Icterus unicolor, *Licht.*
Icterus sulcirostris, *Spix*, Av. Br. pl. 64. f. 2.

This species is common in flocks on the pasture grounds of Chile, and along the whole western shore of the southern part of the continent. In Chile it is called, according to Molina, "cureu." It is a noisy, chattering bird, and runs in the manner of our starlings. It can be taught to speak, and is sometimes kept in cages. It builds in bushes.

MOLOTHRUS NIGER. *Gould.*

Tanagra bonariensis, *Gmel.*
Icterus niger, *Dand.*
Passerina discolor, *Vieill.*
Icterus maxillaris, *D'Orb. & Lafr.*
Icterus sericeus, *Licht.*
Psarocolius sericeus, *Wagl.*

This Molothrus is common in large flocks on the grassy plains of La Plata, and is often mingled with the *Leistes anticus*, and other birds. In the same flock with the usual black kind, there were generally a few dull brown coloured ones, (*Icterus sericeus* of *Licht.*) which I presume are the young. Azara states that the brown-coloured birds are smaller than the black glossy ones, and that they sometimes form one-tenth of the whole number in a flock. In the single specimen which I brought home, the size, with the exception of the length of the wing, is only a very little less. Sonnini, in his notes to Azara, considers the brown birds as the females; I can, however, scarcely believe that so obvious a solution of the difficulty could have escaped so accurate an observer as Azara, These birds in La Plata often may be seen standing on the back of a cow or horse. While perched on a hedge, and pluming themselves in the sun, they sometimes attempt to sing or rather to hiss: the noise is very peculiar; it resembles that of bubbles of air passing rapidly from a small orifice under water, so as to produce an acute sound. Azara states that this bird, like the cuckoo, deposits its eggs in other birds' nests. I was several times told by the country people, that there was some bird which had this habit; and my assistant in collecting, who is a very accurate person, found in the nest of the *Zonotrichia ruficollis* (a bird which occupies in the ornithology of S. America the place of the common sparrow of Europe), one egg larger than the others, and of a different colour and shape. This egg is rather less than that of the missel-thrush, being ·93 of an inch in length, and ·78 in breadth; it is of a bulky form, thick in the middle. The ground colour is a pale pinkish-white, with irregular spots and blotches of a bright reddish-brown, and others less distinct of a greyish hue. This species is evidently a very close analogue of the *M. pecoris* of North America, from which, however it may at once be distinguished by the absence of the glossy brown on the head, neck, and upper breast,—by the metallic blueness of its plumage in the place of a green tinge, and by its somewhat greater size in all its proportions. The young or brown-coloured specimens of these Molothri resemble each other more closely; that of the *M. pecoris* is of a lighter brown, especially under the throat, and the small feathers on its breast and abdomen have each an obscure dark central streak. The eggs of the Molothri, although having the same general character, differ considerably; that of the *M. pecoris* being smaller and less swollen in the middle; it is ·85 of an inch in length, and ·78 in breadth. Its colour cannot be better described than in the words of Dr. Richardson[*]—it is " of a greenish white, with rather small crowded and confluent irregular spots of pale liver-brown, intermixed with others of subdued purplish grey." From this

[*] Fauna Borealis, Birds, p. 278. Dr. Richardson states that the egg is only seven lines and a half in length. I presume the measure of eight lines, instead of twelve to the inch, must in this case have been used. I am much indebted to the kindness of Mr. Yarrell for lending me an egg of the *Molothrus pecoris*, forming part of a collection of North American eggs in his possession.

description it is obvious that the egg of *M. niger* is larger and of a much redder tint; the more prominent spots also are larger, the subdued grey being quite similar in both.

If we were to judge from habits alone, the specific difference between these two species of Molothrus might well be doubted; they seem closely to resemble each other in general habits,—in manner of feeding,—in associating in the same flock with other birds, and even in such peculiarities as often alighting on the backs of cattle. The *M. pecoris*, like the *M. niger*, utters strange noises, which Wilson* describes " as a low spluttering note as if proceeding from the belly." It appears to me very interesting thus to find so close an agreement in structure, and in habits, between allied species coming from opposite parts of a great continent. Mr. Swainson† has remarked that with the exception of the *Molothrus*, the cuckoos are the only birds which can be called truly parasitical; namely, such as " fasten themselves, as it were, on another living animal, whose animal heat brings their young into life, whose food they alone live upon, and whose death would cause theirs during the period of infancy." It is very remarkable, that the cuckoos and the molothri, although opposed to each other in almost every habit, should agree in this strange one of their parasitical propagation: the habit moreover is not universal in the species of either tribe. The Molothrus, like our starling, is eminently sociable, and lives on the open plains without art or disguise:‡ the cuckoo, as every one knows, is a singularly shy bird; it frequents the most retired thickets, and feeds on fruit and caterpillars.§

AMBLYRAMPHUS RUBER. *G. R. Gray.*

Oriolus ruber, *Gmel.*
Amblyramphus bicolor, *Leach.*
Sturnus pyrrhocephalus, *Licht.*
Sturnella rubra, *Vieill.*
Leistes erythrocephala, *Swains.* Class. Birds.

This bird frequented marshy places in the neighbourhood of Maldonado, but it was not common there. It is more solitary than the following allied species; I have, however, seen it in a flock. Seated on a twig, with its beak widely open, it often makes a shrill, but plaintive and agreeable cry, which is sometimes single

---

* Wilson's American Ornithology, vol. ii. p. 162.
† Magazine of Zoology and Botany, vol. i. p. 217.   ‡ See Azara, vol. iii. p. 170.
§ It appears that the eggs in the same nest with that of the *Molothrus pecoris*, are turned out by the parent birds before they are hatched, owing to the egg of the *M. pecoris* being hatched in an unusually short time; in the case of the young cuckoo, as is well known, the young bird itself throws out its foster-brothers. Mr. C. Fox, however, (Silliman's American Journal, vol. xxix. p. 292), relates an instance of three young sparrows having been found alive with a Molothrus.

and sometimes reiterated. Its flight is heavy. The young have their heads and thighs merely mottled with scarlet.

STURNELLA MILITARIS. *Vieill.*

Sturnus militaris, *Gmel.*
Etourneau des terres Magellanique, Pl. enl. 113.

I met with specimens of this bird on the east coast of the continent from the Falkland Islands to 31° S., and on the western coast from the Strait of Magellan to Lima, a space of forty degrees of latitude.

FAMILY.—TROCHILIDÆ.

1. TROCHILUS FLAVIFRONS.

Monte Video.—November. Not abundant.

2. TROCHILUS FORFICATUS. *Lath.*

Edwards' Gleanings.
Vieill. Ois. dores, t. 1.
Ornismya Kingii, *Less.* Trochilidees, pl. 38.

This species is found over a space of 2,500 miles on the west coast, from the hot dry country of Lima to the forests of Terra del Fuego, where it has been described by Captain King as flitting about in a snow-storm. In the wooded island of Chiloe, which has an extremely damp climate, this little bird, skipping from side to side amidst the humid foliage, and uttering its acute chirp, is perhaps more abundant than any other kind. It there very commonly frequents open marshy ground, where a kind of bromelia grows: hovering near the edge of the thick beds, it every now and then dashes in close to the ground; but I could not see whether it ever actually alighted. At that time of the year there were very few flowers, and none whatever near the beds of bromelia. Hence, I was quite sure that they did not live on honey; and on opening the stomach and upper intestine, by the aid of a lens, I could plainly distinguish in a yellow fluid, morsels of the wings of diptera,—probably Tipulidæ. It is evident that these birds search for minute insects in their winter quarters under the thick foliage. I opened the stomachs of several specimens which were shot in different parts of the continent, and in all remains of insects were numerous, forming a black comminuted mass. In one killed at Valparaiso, I found portions of an ant. Amongst the Chonos Islands, at a season when there were flowers in open places, yet the damp recesses of the forests appeared their favourite haunt. In central

Chile these birds are migratory; they make their appearance there in autumn; the first arrival which I observed was on the 14th of April (corresponding to our October) but by the 20th they were numerous. They stay throughout the winter, and begin to disappear in September: on October 12th, in the course of a long walk, I saw only one individual. During the period of their summer migration, nests were very common in Chiloe and the Chonos Island, countries south of Chile. When this species of *Trochilus* migrates southward, it is replaced in Chile by a larger kind, which will be presently described. The migration of the humming birds on both the east[*] and west coasts of North America, exactly corresponds to that which takes place in the southern half of the continent. In both they move towards the tropic during the colder parts of the year, and retreat poleward before the returning heat. Some, however, remain during the whole year in Tierra del Fuego; and in northern California,—which in the northern hemisphere, has this same relative position which Tierra del Fuego has in the southern,—some, according to Beechey, likewise remain. Near the south end of Chiloe, I found on the 8th of December, a nest with eggs nearly hatched. It was of the ordinary form of nests; rather more than an inch in internal diameter, and not deep, composed externally of coarse and fine moss, neatly woven together, and lined with dried confervæ, now forming a very fine reddish fibrous mass. I feel no doubt regarding the nature of this latter substance, as the transverse septa are yet quite distinct: hence this humming bird builds its nest entirely of cryptogamic plants. Egg perfectly white, elongated, or rather almost cylindrical, with rounded ends; length ·557 of an inch, and transverse diameter ·352 of an inch. In January, at the Chonos Islands, when there were young in the nest, a considerable number of old birds were shot; of these, however, few or scarcely any had the shining crest of the male. In the only specimen, which I carefully examined, the metallic tips of the young feathers of the crest, were just beginning to protrude. Several of these males without their crest, had a yellowish gorge; and I saw some with a few light brown feathers on their backs. I presume these appearances are connected with their state of moult.

3. Trochilus Gigas, *Vieill.*

Orsimya tristis, *Less.*, Oiseaux Mouches, pl. 3.

This species is common in central Chile. It is a large bird for the delicate family to which it belongs. At Valparaiso, in the year 1834, I saw several of these birds in the middle of August, and I was informed they had only lately arrived from the parched deserts of the north. Towards the middle of September

---

[*] Humboldt, Pers. Narr. vol. v. part 1. p. 352. Cook's Third Voyage, vol. ii. and Beechey's Voyage.

(the vernal equinox) their numbers were greatly increased. They breed in central Chile, and replace, as I have before said, the foregoing species, which migrates southward for the same purpose. The nest is deep in proportion to its width; externally three inches and a half deep; internal depth a little under one inch and three quarters; width within one inch and two-tenths; mouth slightly contracted. Externally it is formed of fine fibrous grass woven together, and attached by one side and bottom to some thin upright twigs; internally it is thickly lined with a felt, formed of the pappus of some composite flower. When on the wing, the appearance of this bird is singular. Like others of the genus, it moves from place to place, with a rapidity which may be compared to that of Syrphus amongst diptera, and Sphinx among moths; but whilst hovering over a flower, it flaps its wings with a very slow and powerful movement, totally different from that vibratory one common to most of the species, which produces the humming noise. I never saw any other bird, where the force of its wings appeared (as in a butterfly) so powerful in proportion to the weight of its body. When hovering by a flower, its tail is constantly expanded and shut like a fan, the body being kept in a nearly vertical position. This action appears to steady and support the bird, between the slow movements of its wings. Although flying from flower to flower in search of food, its stomach generally contained abundant remains of insects, which, I suspect, are much more the object of its search than honey is. The note of this species, like that of nearly the whole family, is extremely shrill.

In the Appendix an anatomical description of this bird by Mr. Eyton is given.

Order—SCANSORES.

1. Conurus murinus, *Kuhl*.

Psittacus murinus, *Gmel*.
Perruche, *Pernet*, voy. 1. p. 312.

This parrot feeds in large flocks on the grassy plains of Banda Oriental, where not a tree can be seen. They are very destructive to the corn-fields. I was assured that in one year, near Colonia del Sacramiento, on the north bank of the Plata, 2,500 were killed, a reward being given for each dozen heads. Many of these birds build their nests close together in trees, the whole composing a vast mass of sticks. I saw several of their compound nests on the islands in the river Parana.

2. Conurus patachonicus.

Psittacus Patagonus, *Vieill.* Ency. Meth. p.
Psittacara Patagonica, *Less.* Voy. de la Coquille Zool. pl. 35 bis.
Psittacara Patachonica, *Lear's* Ill. Psitt.
Le Patagon, *Azara*, No. 277.
Pattagonian maccaw, *Lath.* Hist. 11, 105.

I obtained specimens of this bird at Bahia Blanca in Northern Patagonia, where there is not a single tree, and the country is dry and very sterile. I did not meet with this species in the southern parts of Patagonia, but it is common near Concepcion in Chile, in nearly the same latitude. They build their nests in holes in cliffs of earth or gravel, together with the *Hirundo cyanoleuca*. In September, at Bahia Blanca, they were laying: their eggs are quite white, and small in proportion to the bird. Several usually rush forth from their holes at the same instant, and utter a noisy scream.

Picus kingii. *G. R. Gray.*

Picus melanocephalus, *King*, Proc. Zool. Soc. 1830, p. 14.

I procured specimens at Valparaiso, and at the Peninsula of Tres Montes (Lat. 46° S.) At the latter place, I killed in January a pair, male and female. Captain King's specimens were obtained from Chiloe. The male has its whole head scarlet with only the nape black, so that Captain King's specific name is unfortunately not applicable for the species; therefore Mr. G. R. Gray thinks it should be named after the first describer. The head of the female is black, with some short reddish-brown feathers over nostrils. There appears to be no other difference in the plumage of the sexes.

Chrysoptilus campestris. *Swains.*

Picus campestris, *Licht.* Cat. p. *Spix*, Av. Br. pl. 116.
Le charpentier des champs, *Azara*, No. 253.

My specimens were obtained from Banda Oriental and Buenos Ayres; I saw it no further southward. Spix says (Birds of Brazil. vol. i. p. 51.) it inhabits Minas Geraes. They frequent open plains and especially rocky ground. They are rather wild, and generally live three or four together. The tail of these ground woodpeckers seems but little used; their beaks, however, were generally muddy to the base: in the stomach of one I found only ants. Their flight is undulatory like that of the English woodpecker, and their loud cry is likewise similar, but each note more separate. They alight on the branch of a tree, horizontally, in the manner of ordinary birds; but occasionally I have seen one clinging in an upright position to a post. They appear to feed exclusively on the ground.

COLAPTES CHILENSIS. *Vigors*.
Picus Chilensis, *Garnot*, Voy. de la Coquille, Zool. pl. 52.

This bird frequents the dry stony hills of central Chile, on which only a few bushes and trees grow. It is closely related in habits and structure to the foregoing species, and appears to be its representative on the western side of the Cordillera; hence I cannot but think the institution of the above two genera unfortunate. It is the "*Pitui*" of Molina, which name, I imagine, it derives from its peculiar cry. Molina states, that it builds its nest in holes in banks.

1. DIPLOPTERUS NÆVIUS. *Boie*.
Cuculus nævius, *Lath*. Ind. 220.

Rio de Janeiro. April.

2. DIPLOPTERUS GUIRA. *G. R. Gray*.
Cuculus guira, *Linn*.
Crotophaga Piririgua, *Vieil*. Gal. des Ois. pl. 44.
Ptiloleptus cristatus, *Swains*.

Buenos Ayres. In small flocks; a noisy, chattering bird.

CROTOPHAGA ANI. *Linn*.
Petit Bout-de-Petun, pl. enl. 102. f. 2.

Rio de Janeiro. May. The stomach of several specimens contained remains of numerous Orthopterous, and some Coleopterous insects.

Order GYRATONES. *Bonap*.
1. COLUMBA FITZROYII. *King*.
Columba Fitzroyii, *King*., in Proc. of Zool. Soc. part 1, 1830, p. 14.
Columba denisea, *Temm*. pl. col. 502.
Columba araucana, *Less*. Voy. de Coqu. pl. 40.?

Peninsula of Tres Montes. Lat. 46° S. January. Captain King's specimens were obtained at Chiloe, three degrees northward. I procured other specimens near Valparaiso. This bird therefore frequents dry rocky land, and damp impervious forests.

2. COLUMBA LORICATA. *Licht.* Vög. Verz. s. 67.

Columba gymnophthalmus, *Temm.*, Pig. i. 18.
─────── leucoptera, *Pr. Max.* Reise, 2, p. 242.
─────── picazuro, *Temm.* Pig. p. 111.
Picazuro, *Azara*, Voy. No. 317.

Frequents in large flocks the fields of Indian corn in the neighbourhood of Maldonado. Legs dull "carmine red." This, probably, is the representative on the eastern side of the Andes of the foregoing or Chilian species.

1. ZENAIDA AURITA. *G. R. Gray.*

Columba aurita, *Temm.* Pig. p. 60. *Wagl.* sp. 70.

I procured specimens of this bird at Maldonado (where it was very abundant) in La Plata, and at Valparaiso in Chile.

2. ZENAIDA GALAPAGOENSIS. *Gould.*

PLATE XLVI.

Z. vertice, cervice, dorso caudæque tegminibus obscurè fuscis vinaceo-tinctis; dorso nigro-guttato; alarum tegminibus fuscis, plumâ singulâ pallidè vinaceo-fusco terminatâ, pogonii utriusque margine, maculâ oblongâ magnâ nigrâ, lineâ albâ separatâ; remigibus primariis et secundariis nigrescenti-fuscis, cinerascenti-albo angustè marginatis; caudâ fuscescenti cinereo ad apicem fasciâ latâ irregulari nigra; loris lineâque angustâ supra et infra oculari nigris pallidè fusco mixtis; gulâ pectoreque vinaceis, colli lateribus ærato tinctis; crisso, caudæque tegminibus inferioribus cinerascentibus, rostro nigro, pedibus rufescenti aurantiacis.

Long. tot. 8½ unc.; *alæ*, 5¼; *caudæ*, 3¼; *tarsi*, ⅞; *rostri*, 1.

Crown of the head and back of the neck, dark chocolate brown, with a vinous tinge; back and tail-coverts the same, the former spotted with black; wing-coverts brown, each feather having a large oblong spot of black on the margin of either web, separated by a line of white, and tipped with light vinous brown, the white predominating on the larger coverts, primaries and secondaries blackish-brown, finely edged with greyish-white; tail brownish-grey, crossed near the extremity with a broad irregular band of black; lores and a narrow line above and beneath the eye black, interrupted with light brown: throat and chest rich vinous, glossed on the sides of the neck with metallic bronze, and fading into greyish on the vent and under tail-coverts; bill black; feet reddish-orange.

Habitat, Galapagos Archipelago. (Sept. and Oct.)

This species may at once be distinguished from the *Z. aurita*, by the redder tint of its breast,—the greater number of black marks on the wing coverts and back—the outer half of some of the feathers on the wing coverts being white—the marks on the under side of the tail being grey (instead of white as in the *Z. aurita*) and by the larger size of its beak.

This dove is one of the most abundant birds in the Archipelago. It frequents the dry rocky soil of the low country, and often feeds in the same flock with the several species of *Geospiza*. It is exceedingly tame, and may be killed in numbers. Formerly it appears to have been much tamer than at present. Cowley,[*] in 1684, says that the "Turtle doves were so tame that they would often alight upon our hats and arms, so as that we could take them alive: they not fearing man, until such time as some of our company did fire at them, whereby they were rendered more shy." Dampier[†] (in the same year) also says that a man in a morning's walk might kill six or seven dozen of these birds. At the present time, although certainly very tame, they do not alight on people's arms; nor do they suffer themselves to be killed in such numbers. It is surprising that the change has not been greater;—for these islands during the last hundred and fifty years, have been frequented by buccaneers and whalers; and the sailors, wandering through the woods in search of tortoises, take delight in knocking down the little birds.

3. ZENAIDA BOLIVIANA. *G. R. Gray.*

Columba Boliviana, *D'Orb. & Lafr.* Mag. de Zool. 1836. Ois. p. 33. pl. 75.

My specimen was obtained (end of August) at Valparaiso.

1. COLUMBINA STREPITANS. *Spix.*

(Av. pl. 75, f. 1.)

I procured specimens at Maldonado (where it was not common), on the banks of the Plata, and at Rio Negro, in Northern Patagonia.

2. COLUMBINA TALPACOTI. *G. R. Gray.*

Columba Talpacoti, *Temm.* Pig. p. 22. t. 12.
Columbina Cabocolo, *Spix,* Av. pl. 75a. f. 1.
Le Pigeon rougeatre, *Azara,* No. 323.

My specimens were obtained at Rio de Janeiro.

---

[*] Cowley's Voyage, p. 10, in Dampier's Collection of Voyages.

[†] Dampier's Voyage, vol. i. p. 103. For some further observations on the tameness of the birds on this and some other islands, see my Journal of Researches, p. 475.

1. Attagis Falklandica. *G. R. Gray.*

Tetrao Falklandicus, *Gmelin*, Syst. 1. 762.
La Caille des Isles Malouines, *Buff.* pl. enl. 222.
Coturnix Falklandica, *Bonn.* Ency. Meth. Orn. 220.
Perdix Falklandica, *Lath.* Ind. Orn. 11, 652.
Ortyx Falklandica, *Steph.* Shaw's Zool. xi. 386.

This bird is not uncommon on the mountains in the extreme southern parts of Tierra del Fuego. It frequents, either in pairs or small coveys, the zone of alpine plants above the region of forest. It is not very wild, and lies very close on the bare ground.

2. Attagis gayii. *Less.*

Attagis Gayii, *Less.* Cent. Zool. pl. 47, p. 155.

A specimen was given me, which was shot on the lofty Cordillera of Coquimbo, only a little below the snow-line. At a similar height, on the Andes, behind Copiapo, which appear so entirely destitute of vegetation, that any one would have thought that no living creature could have found subsistence there, I saw a covey. Five birds rose together, and uttered noisy cries; they flew like grouse, and were very wild. I was told that this species never descends to the lower Cordillera. These two species, in their respective countries, occupy the place of the ptarmigan of the northern hemisphere.

Tinochorus rumicivorus. *Eschsch.*

Thinocorus rumicivorus, *Eschsch.* Zool. Atl. pl. 2.
Tinochorus Eschscholtzii, *Less.* Cent. Zool. pl. 50.

This very singular bird, which in its habits and appearance partakes of the character both of a wader and one of the gallinaceous order, is found wherever there are sterile plains, or open dry pasture land, in southern South America. We saw it as far south as the inland plains of Patagonia at Santa Cruz, in lat. 50°. On the western side of the Cordillera, near Concepcion, where the forest land changes into an open country, I saw this bird, but did not procure a specimen of it: from that point throughout Chile, as far as Copiapo, it frequents the most desolate places, where scarcely another living creature can exist: it thus ranges over at least twenty-three degrees of latitude. It is found either in pairs or in small flocks of five or six; but near the Sierra Ventana I saw as many as thirty and forty together. Upon being approached they lie close, and then are very difficult to be distinguished from the ground; so that they often rise quite unexpectedly. When feeding they walk rather slowly, with their legs wide apart. They dust themselves in roads and sandy places. They frequent particular spots, and may be found there day after day. When a pair are together, if one is shot, the other seldom rises; for these birds, like partridges, only take wing in a flock. In all these respects, in the muscular gizzard adapted for vegetable food, in the arched beak and fleshy nostrils, short legs, and form of foot, the Tinochorus has a close affinity with quails. But directly the bird is seen flying, one's opinion is changed; the long pointed wings, so different from those in the gallinaceous order, the high irregular flight, and plaintive cry uttered at the moment of rising, recall the idea of a snipe. Occasionally they soar like partridges when on the wing in a flock. The sportsmen of the Beagle unanimously called it the short-billed snipe. To this genus, or rather to that of the sandpiper, it approaches, as Mr. Gould informs me, in the shape of its wing, the length of the scapulars, the form of the tail, which closely resembles that of *Tringa hypoleucos*, and in the general colour of the plumage. The male bird, however, has a black mark on its breast, in the form of a yoke, which may be compared to the red horseshoe on the breast of the English partridge. Its nest is said to be placed on the borders of lakes, although the bird itself is an inhabitant of the parched desert. I was told that the female lays five or six white eggs, spotted with red. I opened the stomachs of many specimens at Maldonado, and found only vegetable matter, which consisted of chopped pieces of a thick rushy grass, and leaves of some plant, mixed with grains of quartz. The contents of the intestine and the dung were of a very bright green colour. At another season of the year, and further south, I found the craw of one full of small seeds and a single ant. Those which I shot were exceedingly fat, and had a strong offensive game odour; but they are said to be very good eating, when cooked. Pointers will stand to them. In the Appendix Mr. Eyton has given an anatomical description of this bird, which partly confirms that affinity both to the Grallatores and Razores, which is so remarkable in its habits and external appearance.

CHIONIS ALBA. *Forst.*
Shaw's Nat. Miscel. pl. 481.

I opened the stomach of a specimen killed at the Falkland Islands, and found in it small shells, chiefly Patellæ, pieces of sea-weed, and several pebbles. The contents of the stomach and body smelt most offensively. Forster remarked this circumstance; but since his time, other observers, namely, Anderson, Quoy, Gaimard, and Lesson (Manuel d'Ornithologie, tom ii, p. 342) have found that this is not always the case, and they state that they have actually eaten the Chionis. I was not aware of these observations, but independently was much surprised at the extraordinary odour exhaled. We, like other voyagers in the Antarctic seas, were struck at the great distance from land, at which this bird is found in the open ocean. Its feet are not webbed, its flight is not like that of other pelagic birds, and the contents of its stomach, and structure of legs, show that it is a coast-feeder. Does it frequent the floating icebergs of the Antarctic ocean, on which sea-weed and other refuse is sometimes cast?

1. NOTHURA MAJOR. *Wagl.*

Nothura major, *Wagl.* Syst. Av. p.   sp. 4.
Tinamus major, *Spix.* Av. pl. 80.

These birds are very common on the northern shores of the Plata. They do not rise in coveys, but generally by pairs. They do not conceal themselves nearly so closely as the English partridge, and hence great numbers may be seen in riding across the open grassy plains. Note, a shrill whistle. It appears a very silly bird: a man on horseback, by riding round and round in a circle, or rather in a spire, so as to approach closer each time, may knock on the head almost as many as he pleases. The more common method is to catch them with a running noose, or little lazo, made of the stem of an ostrich's feather, fastened to the end of a long stick.* A boy on a quiet old horse will frequently thus catch thirty or forty in a day. The flesh of this bird, when cooked, is most delicately white, but rather tasteless.

The egg of this species, I believe, closely resembles that of the two following.

2. NOTHURA MINOR. *Wagl.*

Nothura minor, *Wagl.* Syst. Av. p.   sp. 4.
Tinamus minor, *Spix,* Av. Br. pl. 82.

I procured a specimen of this bird at Bahia Blanca, in northern Patagonia, where it frequented the sand-dunes and the surrounding sterile plains. Its habits appear similar to those of the *N. major,* but it lies closer and does not so readily take to the wing. It is the smallest of the species mentioned in this work, and its plumage is less distinctly spotted. The egg of this bird is described below. Spix's specimens were obtained at Tijuco in Brazil. The figure in his work on the Birds of Brazil, differs slightly from mine, in being less marked on the breast.

3. NOTHURA PERDICARIA *G. R. Gray.*

Crypturus perdicarius, *Kittlitz,* Vögel von Chili.

This species closely resembles, in its general appearance and habits, the

---

* In Hearne's Travels in North America, (p. 383), it is stated that the Northern Indians shoot the varying hare, which will not bear to be approached in a straight line, in an analogous manner, by walking round it in a spire. The middle of the day is the best time, when the shadow of the hunter is not very long.

N. major, of which probably it is the analogue on the western side of the Cordillera. It is larger and has a considerably longer beak than the *N. major*; its breast is not spotted, and its abdomen has a less fulvous tinge. The *N. perdicarius* runs on the open ground, generally a pair together, in the same unconcealed manner, as its analogue, and does not readily lie close. Flight similar, but on rising it utters a shriller whistle, of a different tone. It does not appear to be so easily caught as the Plata species. It is tolerably abundant in all parts of Chile, as far north as the valley of Guasco; but I was assured, that it has never been seen in the valley of Copiapo, although only seventy miles north of Guasco, and of a similar character. The egg is very glossy and of a peculiar colour, which, according to Werner's nomenclature, is a palish chocolate red: length in longer axis 2·07 of an inch; shorter axis 1·495 of an inch. The egg of the *N. minor* is of a similar colour, but a shade paler, and rather smaller; its length being 1·815, and its transverse diameter 1·3 of an inch.

RHYNCHOTUS RUFESCENS. *Wagl.*

Rhynchotus rufescens, *Wagl.* Av. Syst.
Tinamus rufescens. *Temm.* Gall. iii. p. 552.
Rhynchotus fasciatus. *Spix.* Av. Br. pl. 76.
Cryptura Guaza. *Vieill.*
Crypturus rufescens. *Licht.* Vög. Verz. s. 67.

My specimens were procured at Maldonado, where it is a much rarer bird than the *Nothura major;* I met with it also in the sterile country near Bahia Blanca. At Maldonado it frequented swampy thickets on the borders of lakes. It lies very close, and is unwilling to rise, but often utters, whilst on the ground, a very shrill whistle. When on the wing, it flies to a considerable distance. Several are generally found together, but they do not rise at the same instant, like a covey of partridges. Flesh, when cooked, perfectly white. Spix's specimens were procured in the country between St. Paul's and Minas Geraes; so that this bird, as well as the *Nothura minor*, has a considerable range.

ORDER—CURSORES. *Temm.*

1. RHEA AMERICANA. *Lath.*

This bird is well known to abound on the plains of La Plata. To the north it is found, according to Azara, in Paraguay, where, however, it is not common; to the south its limit appears to be from 42° to 43°. It has not crossed the Cordillera; but

I have seen it within the first range of mountains on the Uspallata plain, elevated between six and seven thousand feet. The ordinary habits of the ostrich are well known. They feed on vegetable matter, such as roots and grass; but at Bahia Blanca, I have repeatedly seen three or four come down at low water to the extensive mud-banks which are then dry, for the sake, as the Gauchos say, of catching small fish. Although the ostrich in its habits is so shy, wary, and solitary, and although so fleet in its pace, it falls a prey, without much difficulty, to the Indian or Gaucho armed with the bolas. When several horsemen appear in a semicircle, it becomes confounded, and does not know which way to escape. They generally prefer running against the wind; yet at the first start they expand their wings, and like a vessel make all sail. On one fine hot day I saw several ostriches enter a bed of tall rushes, where they squatted concealed, till quite closely approached. It is not generally known that ostriches readily take to the water. Mr. King informs me that in Patagonia, at the Bay of San Blas and at Port Valdes, he saw these birds swimming several times from island to island. They ran into the water, both when driven down to a point, and likewise of their own accord, when not frightened: the distance crossed was about 200 yards. When swimming, very little of their bodies appear above water, and their necks are extended a little forward: their progress is slow. On two occasions, I saw some ostriches swimming across the Santa Cruz river, where it was about four hundred yards wide, and the stream rapid. Captain Sturt,* when descending the Murrumbidgee, in Australia, saw two emus in the act of swimming.

The inhabitants who live in the country readily distinguish, even at a distance, the male bird from the female. The former is larger and darker coloured,† and has a larger head. The ostrich, I believe the cock, emits a singular, deep-toned, hissing note. When first I heard it, standing in the midst of some sand-hillocks, I thought it was made by some wild beast, for it is a sound that one cannot tell whence it comes, or from how far distant. When we were at Bahia Blanca in the months of September and October, the eggs were found, in extraordinary numbers, all over the country. They either lie scattered single, in which case they are never hatched, and are called by the Spaniards, huachos, or they are collected together into a shallow excavation, which forms the nest. Out of the four nests which I saw, three contained twenty-two eggs each, and the fourth twenty-seven. In one day's hunting on horseback sixty-four eggs were found; forty-four of these were in two nests, and the remaining twenty scattered huachos. The Gauchos unanimously affirm, and there is no reason to doubt their statement, that the male

---

* Sturt's Travels, vol. ii, p. 74.
† A Gaucho assured me that he had once seen a snow-white, or Albino variety, and that it was a most beautiful bird.

bird alone hatches the eggs, and for some time afterwards accompanies the young. The cock when on the nest lies very close; I have myself almost ridden over one. It is asserted that at such times they are occasionally fierce, and even dangerous, and that they have been known to attack a man on horseback, trying to kick and leap on him. My informer pointed out to me an old man, whom he had seen much terrified by one chasing him. I observe, in Burchell's Travels in South Africa, that he remarks, "having killed a male ostrich, and the feathers being dirty, it was said by the Hottentots to be a nest bird." I understand that the male emu, in the Zoological Gardens, takes care of the nest: this habit therefore is common to the family.*

The Gauchos unanimously affirm that several females lay in one nest. I have been positively told, that four or five hen birds have been actually watched and seen to go, in the middle of the day, one after the other, to the same nest. I may add, also, that it is believed in Africa, that two or more females lay in one nest.† Although this habit at first appears very strange, I think the cause may be explained in a simple manner. The number of eggs in the nest varies from twenty to forty, and even to fifty; and according to Azara to seventy or eighty. Now although it is most probable, from the number of eggs found in one district being so extraordinarily great, in proportion to that of the parent birds, and likewise from the state of the ovarium of the hen, that she may in the course of the season lay a large number, yet the time required must be very long. Azara states,‡ that a female in a state of domestication laid seventeen eggs, each at the interval of three days one from another. If the hen were obliged to hatch her own eggs, before the last was laid, the first probably would be addled; but if each laid a few eggs at successive periods, in different nests, and several hens, as is stated to be the case, combined together, then the eggs in one collection would be nearly of the same age. If the number of eggs in one of these nests is, as I believe, not greater on an average than the number laid by one female in the season, then there must be as many nests as females, and each cock bird will have its fair share of the labour of incubation; and this during a period when the females probably could not sit, on account of not having finished laying.§ I have before mentioned the great numbers of huachos, or scattered

---

* It appears, also, from Mr. Gould's late most interesting discoveries regarding the habits of the *Talegalla Lathami*, (an Australian bird, one of the Rasores,) that several females lay in one nest, and that the eggs are hatched by the heat engendered by a mass of decaying vegetable matter. It appears that the males assist the females in scratching together the leaves and earth, of which the great conical mound or nest is composed.

† Burchell's Travels, vol. i. p. 280.   ‡ Azara, vol. iv. p. 173.

§ Lichtenstein, however, (Travels, vol. ii. p. 25,) states, that the hens begin to sit when ten or twelve eggs are laid, and that they afterwards continue laying. He affirms that by day the hens take turns in sitting, but that the cock sits all night.

eggs; so that in one day's hunting the third part found were in this state. It appears odd that so many should be wasted. Does it not arise from some difficulty in several females associating together, and in finding a male ready to undertake the office of incubation? It is evident that there must at first be some degree of association, between at least two females; otherwise the eggs would remain scattered at distances far too great to allow of the male collecting them into one nest. Some authors believe that the scattered eggs are deposited for the young birds to feed on. This can hardly be the case in America, because the huachos, although often found addled and putrid, are generally whole.

2. Rhea Darwinii. *Gould.*

Plate XLVII.

Gould, in Proceedings of Zoological Soc. 1837, p. 35.

R. pallide fusca, plumâ singulâ distinctâ semilunari notâ candidâ terminatâ; capite collo, femoribusque pallidioribus: rostri culmine augusti, ad apicem latiore, frontes plumis parvis setosis anticè directis et supra nares extensis; tarsi lateribus in dimidiam partem plumis parvis mollibus tectis; tarso ⅔ antice posticeque toto, squamis reticulatis tecto.

Long. tot. 52 unc.; *alæ*, 30; *tarsi*, 11; *rostri*, 2.

The whole of the plumage light brown, each feather with a decided crescent-shaped mark of pure white at the extremity; head, neck, and thighs lighter; base of the neck blackish; culmen of the bill narrow, becoming a little broader towards apex; front with small bristly feathers, pointing forwards and reaching over the nostrils. Tarsus with small downy feathers on sides, extending half way downwards; upper two-thirds of front of tarsus, and whole hinder side, with reticulated scales.

Habitat, Eastern Patagonia (Lat. 40° S. to 54° S.)

This species, which Mr. Gould, in briefly characterizing it at a meeting of the Zoological Society, has done me the honour of calling after my name, differs in many respects from the *Rhea Americana*. It is smaller, and the general tinge of the plumage is a light brown in place of grey; each feather being conspicuously tipped with white. The bill is considerably smaller, and especially less broad at its base; the culmen is less than half as wide, and becomes slightly broader towards the apex, whereas in the *R. Americana* it becomes slightly narrower; the extremity, however, of both the upper and the lower mandible, is more tumid in the latter, than in the *R. Darwinii*.

	R. Darwinii.	R. Americana.
	inches	inches
Length of beak, from edge of membrane at base to the apex	2	$2\frac{2}{8}$
Length, from anterior margin of eye to apex	$3\frac{4}{12}$	$5\frac{5}{12}$
Width of upper mandible, measured across middle of nostrils	$1\frac{1}{20}$	$1\frac{3}{20}$

The skin round and in front of the eyes is less bare in *R. Darwinii;* and small bristly feathers, directed forwards, reach over the nostrils. The feet and tarsi are nearly of the same size in the two species. In the *R. Darwinii,* short plumose feathers extend downwards in a point on the sides of the tarsus, for about half its length. The upper two-thirds of the tarsus, in front, is covered with reticulated scales in place of the broad transverse band-like scales of the *R. Americana;* and the scales of the lower third are not so large as in the latter. In the *R. Darwinii* the entire length of the back of the tarsus is covered with reticulated scales, which increase in size from the heel upwards: in the common *Rhea,* the scales on the hinder side of the tarsus are reticulated only on the heel, and about an inch above it; all the upper part consisting of transverse bands, similar to those in front.

The first notice I received of this species was at the Rio Negro, in Northern Patagonia, where I repeatedly heard the Gauchos talking of a very rare bird, called *Avestruz Petise.* They described it as being less than the common ostrich (which is there abundant), but with a very close general resemblance. They said its colour was dark and mottled, and that its legs were shorter, and feathered lower down than those of the common ostrich. It is more easily caught by the bolas than the other species. The few inhabitants who had seen both kinds, affirmed that they could distinguish them apart, from a long distance. The eggs, however, of the small species appeared more generally known, and it was remarked with surprise, that they were very little less than those of the common *Rhea,* but of a slightly different form, and with a tinge of pale blue. Some eggs which I picked up on the plains of Patagonia, agree pretty well with this description; and I do not doubt are those of the Petise. This species occurs most rarely in the neighbourhood of the Rio Negro; but about a degree and a half further south they are tolerably abundant. One Gaucho, however, told me he distinctly recollected having seen one, many years before, near the mouth of the Rio Colorado, which is north of the Rio Negro. They are said to prefer the plains near the sea. When at Port Desire in Patagonia (Lat. 48°), Mr. Martens shot an ostrich; I looked at it, and from most unfortunately forgetting at the moment, the whole subject of the Petises, thought it was a two-third grown one of the common sort. The bird was skinned and cooked before my memory returned. But the head, neck, legs, wings, many of the larger feathers, and a large part of the skin, had been preserved. From these a very nearly perfect specimen has been put together, and is now exhibited in the museum of the Zoological Society. M. A. D'Orbigny, a distinguished French naturalist, when at the Rio Negro, made great exertions to procure this bird, but had not the good fortune to succeed. He mentions it in his Travels (vol. ii. p. 76.) and proposes (in case, I presume, of his obtaining a specimen at some future time, and thus being able to characterize it,) to call it *Rhea pennata*. A notice of this species was given long since (A.D. 1749) by Dobrizhoffer, in his account of the Abipones (vol. i. Eng. Trans. p. 314). He says, " You must know, moreover, that Emus differ in size and habits in different tracts of land; for those that inhabit the plains of Buenos Ayres and Tucuman are larger, and have black, white, and grey feathers; those near to the Strait of Magellan are smaller, and more beautiful, for their white feathers are tipped with black at the extremity, and their black ones in like manner terminate in white."

Among the Patagonian Indians in the Strait of Magellan, we found a half-bred Indian, who had lived some years with this tribe, but had been born in the northern provinces. I asked him if he had ever heard of the Avestruz Petise? He answered by saying, " Why there are none others in these southern countries." He informed me that the number of eggs in the nest of the Petise is considerably less than with the other kind, namely, not more than fifteen on an average; but he asserted that more than one female deposited them. At Santa Cruz we saw several of these birds. They were excessively wary: I think they could see a person approaching, when he was so far off as not to distinguish the ostrich. In ascending the river few were seen; but in our quiet and rapid descent, many, in pairs and by fours or fives, were observed. It was remarked by some of the officers, and I think with truth, that this bird did not expand its wings, when first starting at full speed, after the manner of the northern kind. The fact of these ostriches swimming across the river has been mentioned. In conclusion, I may repeat that the *R. Americana* inhabits the eastern plains of S. America as far as a little south of the Rio Negro, in lat. 41°, and that the *R. Darwinii* takes its place in Southern Patagonia; the part about the Rio Negro being neutral territory. Wallis saw ostriches at Bachelor's river (lat 53° 54'), in the Strait of Magellan, which must be the extreme southern possible range of the Petise.

Order—GRALLATORES.

Oreophilus totanirostris. *Jard. & Selb.*
<small>Oreophilus totanirostris, *Jard. & Selb*. Illustr. of Orn. iii. pl. 151.</small>

My specimens were obtained at Maldonado and at Valparaiso. At the former, it was common, feeding on the open grassy plains in small flocks, mingled with the icteri and the thrush-like *Xolmis variegata*. When these birds rise on the wing, they utter a plaintive cry. Legs "crimson red;" toes leaden colour, with their under surface remarkably soft and fleshy. Iris dark brown.

Charadrius virgininus. *Borkh.*

Charadrius virgininus, *Borkh.* Act. Acad. Cæs. Leop. Car. Nat. Cur. 1834. xvi. pl. 18.
Charadrius marmoratus, *Wagl.*

This representative of the golden plover of Europe and North America, is common on the banks of the Plata in large and small flocks. It is found also, according to Meyer, in Chile.

1. Squatarola cincta. *Jard. & Selby.*

Tringa Urvillii, *Garnot*, Ann. Ic. Nat. Jan. 1826.
Vanellus cinctus, *Less.* Voy. de la Coqu. Zool. p. 720. pl. xliii.
Squatarola cincta, *Jard. & Selby's* Illust. Orn. pl. 110.
Charadrius rubecola, *Vig.* Journ. iv. p. 96.

I obtained specimens of this bird in Tierra del Fuego, where it inhabited both the sea shore and the bare stony summits of the mountains; at the Falkland Islands, where it frequented the upland marshes; and at Chiloe, where I met with large flocks in the fields, not near the coast.

2. Squatarola fusca. *Gould.*

S. vertice corporeque supra fuscis, dorsi parapterique plumis pallidiore marginatis; remigibus primariis nigrescenti fuscis, pogoniis externis albo angustè marginatis rhachibus albis; uropygio caudáque obscurè fuscis, remigibus externis albo latè marginatis et terminatis; fronte, genis, gulâ, abdomine postico, caudæque tegminibus inferioribus flavescenti albis, colli pectorisque lateribus fuscis, colli plumis fusco pallido terminatis; pedibus nigris.

Long. tot. 8 unc. *alæ*, 5⅜; *caudæ*, 3; *tarsi*, 1¾; *rostri*, ¾.

Crown of the head, all the upper surface brown, the feathers of the back and the scapularies, margined with paler; primaries blackish brown, finely edged on their inner margins with white, and with white shafts; rump and tail dark brown, the outer feathers largely margined and tipped with white; forehead and sides of the face sandy white; throat, lower part of the abdomen, and under tail coverts, buffy white; sides of the neck and chest brown; the feathers of the latter tipped with still lighter brown; bill and feet black.

Habitat, Maldonado; inland glassy plains.

This species is most closely allied to the foregoing. I obtained only one specimen, which, on comparison with several of the *S. cincta*, appears a little larger in all its dimensions, especially in the length of the tarsi. Its back and scapularies are of a more uniform brown, the feathers being less edged with pale brown. Its feet are black, whereas those of *S. cincta* are brown.

Philomachus Cayanus. *G. R. Gray.*
Charadrius Cayanus, *Lath.* Ind. Orn. 11. 748.

I met with this bird from latitude 30° to 45° S. on both sides of S. America. In La Plata it is called "Teru-tero," in imitation of its cry; and in Chile, according to Molina, "Theghel." These birds, which in many respects resemble in habits our peewits (*Vanellus cristatus*), frequent, generally in pairs, open grassy land, and especially the neighbourhood of lakes. As the peewit takes its name from the sound of its voice, so does the teru-tero. While riding over the grassy plains, one is constantly pursued by these birds, which appear to hate mankind, and I am sure deserve to be hated, for their never-ceasing, unvaried, harsh screams. The stillness of the night is often disturbed by them. To the sportsman they are most annoying, by announcing to every other bird and animal his approach: to the traveller in the country, they may possibly, as Molina says, do good, by warning him of the midnight robber. During the breeding season, they attempt, like our peewits, by feigning to be wounded, to draw away from their nests dogs and other enemies. Their eggs are of a pointed oval form; of a brownish olive colour, thickly spotted with dark brown. Their eggs, like those of the peewit, are esteemed particularly good eating.

1. Hiaticula Azaræ. *G. R. Gray.*
Charadrius Azaræ, *Temm.* pl. col. 184.
———— collaris, *Vieill.*
Albatuitui à collier noir, *Azara*, No. 392.

My specimens were obtained on the banks of the Plata and at Valparaiso. The specimen from the latter country differs from those procured at the former, in the absence of the black collar on the breast, of the black streak running from the eye to the corner of the mouth; in the plumage of the back and back of head having a lesser tinge of red; and especially in the feet being black, and tarsi blackish, instead of both being orange, as is the case with those killed on the shores of the Plata. I have not, however, thought it desirable to make two species of these birds, not having a larger series of specimens for comparison.

2. Hiaticula trifasciatus. *G. R. Gray.*
Charadrius bifasciatus, *Licht.* Vog. Verz. p. 71.
———— trifasciatus, *Wagl.* Syst. Av. sp. 31.

I procured two specimens of this bird at Bahia Blanca, in Northern Patagonia.

3. HIATICULA SEMIPALMATA. *G. R. Gray.*
Tringa semipalmata, *Temm.*
Charadrius semipalmatus, *Caup.* Isis. 1825, p. 1375, t. 14. *Wagl.* Syst. Av. sp. 23.
Bonap. Am. Orn. iv. pl. 25, f. 4.

Galapagos Archipelago.

HÆMATOPUS PALLIATUS. *Temm.*

Rio Plata.

EGRETTA LEUCE. *Bonap.*
Ardea Leuce, Ill.
Ardea Egretta, *Wils.* Am. Orn. pl. 61, f. 4.

My specimen was procured at Maldonado. I saw it also in Patagonia.

ARDEA HERODIAS. *Linn.*

Galapagos Archipelago. Frequents the sea-coast and salt-lagoons. There are no fresh water pools in any of these islands.

1. NYCTICORAX VIOLACEUS. *Bonap.*
Ardea violacea, *Linn.*
Ardea callocephala, *Wagl.* Syst. Av.

Mr. G. R. Gray has thought it advisable to give the following description of this specimen, from the Gallapagos Archipelago. It appears to be a young bird, and is small in all its dimensions.

Upper part blackish-grey; each feather marked down the middle with a broad stripe of black, and tinged on the margins with shining bronze-brown; beneath the body blueish-grey, with the front of the neck, top of the head, and margins of the feathers on the thighs rufous; the sides of the head and throat deep black, the former divided in the middle on each side with a patch of white; the bill black, and feet of a pale reddish colour.

2. NYCTICORAX AMERICANUS. *Bonap.*
Ardea nycticorax, *Wils.* (young bird.)

Valparaiso, Chile.

THERISTICUS MELANOPS. *Wagl.*
Ibis melanops, *Lath.* Hist. ix. pl. 150.

This bird frequents the desert gravelly plains of Patagonia, as far south as lat. 48°: in the British Museum there are specimens which Captain Clapperton brought from central Africa; so that this bird has an extraordinarily wide range. It generally lives in pairs, but during part of the year in small flocks. Its cry is very singular and loud: when it is heard at a distance it closely resembles the neighing of the guanaco. I opened the stomach of two specimens, and found in them remains of lizards, cicadæ, and scorpions. It builds in rocky cliffs on the sea-shore: egg dirty white, freckled with pale reddish-brown; its circumference over longer axis is seven inches. The legs are carmine and scarlet-red: iris scarlet-red.

Ibis (falcinellus) Ordi. *Bonap.*

Tantalus Mexicanus, *Ord.* Journ. Acad. Phil.
Tantalus chalcopterus? *Temm.*
Ibis Falcinellus, *Bonap.* Am. Orn. iii.

My specimen was obtained at the Rio Negro: it is very numerous in large flocks on the vast swampy plains between Bahia Blanca and Buenos Ayres. Its flight when soaring is singularly graceful; the whole flock moving in precise concert.

1. Numenius hudsonicus. *Lath.*

Numenius Hudsonicus, *Lath.* Ind. Orn. ii. 712.

This curlew is very abundant on the tidal mud-banks of Chiloe. When the flock rises, each bird utters a shrill note.

2. Numenius brevirostris. *Licht.*

Numenius brevirostris, *Licht.* Cat. 75, sp. 774 a.

Buenos Ayres.

Limosa Hudsonica. *Swains.*

Scolopax Hudsonica, *Lath.* Ind. Orn. ii. 720.

My specimens were obtained from the Falkland Islands and from Chiloe, where it frequented the tidal mud-banks in flocks.

1. Totanus flavipes. *Vieill.*

Totanus flavipes, *Vieill.* Ency. Meth. 1106.
Yellow shanks snipe, *Penn.* Arct. Zool. ii. 468.
——————— *Wils.* Am. Orn. pl. 58. f. 4.

Monte Video, Rio Plata.

2. Totanus macropterus. *G. R. Gray.*

Tringa macroptera, *Spix.* Av. n. sp. pl. 92.

Monte Video, Rio Plata.

3. TOTANUS MELANOLEUCOS. *Licht. et Vieill.*

Scolopax melanoleuca, *Gmel.*
Scolopax vociferus, *Wils.* Am. Orn. pl. 58, f. 5.
Chorlito à croupion blanc, *Azara*, No. 394.
Totanus solitarius, *Vieill.*
White-rumped snipe, *Lath.*

Maldonado, Rio Plata.

4. TOTANUS FULIGINOSUS. *Gould.*

T. corpore supra caudâque fuliginoso-griseis; alis fuscis; gutture albo; pectore hypochondriisque plumbeo-griseis; abdomine medio, caudæ tegminibus inferioribus albis, illis obscure, his plane grisescenti fusco fasciatis; rostri rubescenti fusco; pedibus obscurè olivaceo fuscis.

Long. tot. $9\frac{1}{2}$ unc. *alæ*, $6\frac{5}{8}$; *caudæ*, 3; *tarsi*, $1\frac{1}{4}$; *rostri*, $\frac{5}{8}$.

The whole of the upper surface and tail sooty-grey; wings dull brown; throat white; chest and flanks leaden grey; centre of the abdomen and under tail coverts white, the former indistinctly, and the latter distinctly, barred with greyish brown; bill, reddish-brown; feet, dark olive-brown.

Habitat, Galapagos Archipelago (*October*).

This species appear quite distinct from any described one.

HIMANTOPUS NIGRICOLLIS. *Vieill.*

Himantopus nigricollis, *Vieill.* Ency. Meth. 340.
Recurvirostra himantopus, *Wils.* Am. Orn. pl. 58. f. 2.

My specimens were obtained from the provinces bordering the Plata. On the great swampy plains and fens which lie between Buenos Ayres and Bahia Blanca, it is very numerous in small, and occasionally, in large flocks. This plover, which appears as if mounted on stilts, has been wrongfully accused of inelegance; when wading about in shallow water, which is its favourite resort, its gait is far from awkward. In a flock it utters a noise, which singularly resembles the cry of a pack of small dogs in full chase: when I travelled across the above mentioned plains, I was more than once startled, when lying awake at night, at the distant sound, and thought the wild Indians were coming.

TRINGA RUFESCENS. *Vieill.*

Tringa rufescens,*Vieill.*, N. Dict. d'Hist. Nat. 34. p. 470.
—————————Ency. Meth. Orn. p. 1090.
—————————Gal. des Ois. pl. 238.
—————————*Yarrel*, Lin. Trans.
—————————*Gould*; Birds of Europe, pl.

Monte Video, Rio Plata.

1. **Pelidna Schinzii.** *Bonap.*

Tringa Schinzii, *Brehm. Bonap.* Am. Orn. iv. pl. 24. f. 2.
Pelidna cinclus, var. *Say.*

Flocks of this species were common on the shores of the inland bays in the southern parts of Tierra del Fuego.

2. **Pelidna minutilla.** *Gould.*

Tringa minutilla, *Vieill.* Ency. Meth. 1089.

Galapagos Archipelago. Both the specimens which I procured here are smaller than the ordinary size of this bird, but do not differ in other respects. Vieillot says it ranges from the Antilles to Canada.

Rhynchæa semicollaris. *G. R. Gray.*

Totanus semicollaris, *Vieill.*, Ency. Meth. p. 1100.
Rhynchæa Hilairea, *Valenc. Less.* Ill. de Zool. pl. 18.
Rhynchæa occidentalis, *King*, Zool. Journ. iv. 94.
Le chorlito a demi colliers blanc et noiratre, *Azara*, No. 409.

Monte Video, Rio Plata. Frequents swamps; habits like the Scolopax Gallinago.

1. **Scolopax (Telmatias) Paraguaiæ.** *Vieill.*

Scolopax Paraguai, *Veiell.* Ency. Meth. p. 1160.
———Brasiliensis, *Swains.* Faun. Bor. Am. Birds, p. 400.
Bocassine 1st Espece, *Azara.*

Valparaiso and Maldonado, Rio Plata.

2. **Scolopax (Telmetias) Magellanicus.** *King.*

Scolopax Magellanicus, *King*, Zool. Journ.

My specimens were obtained from Maldonado and East Falkland Island. Flight a very little less irregular and rapid than the English snipe. I several times in May observed this, as well as the foregoing species, flying in lofty circles, and suddenly stooping downwards, at the same time that it uttered a peculiar drumming noise, similar to that made by the English snipe in summer, when breeding. This species is most closely allied to the foregoing, but I have no doubt it is distinct; because at the time when I procured specimens of both at Maldonado, I perceived a difference between them. This species is there more abundant than the *S. Paraguaiæ*. Its beak is nearly three-tenths of an inch shorter, and the culmen rather broader. The plumage of its back is of a decidedly less dark tint; each separate feather having much less black in it.

STREPSILAS INTERPRES. *Ill.*

Tringa Morinellus, *L.*

I obtained specimens from Iquique, on the coast of Peru, and from the Galapagos Archipelago.

CREX LATERALIS. *Licht.*

Crex lateralis, *Licht.*, Cat. p.
———— *Griff.* An. King. Aves.

Maldonado, Rio Plata. On being disturbed readily takes wing. Base of the bill, especially of the lower mandible, bright green.

1. ZAPORNIA NOTATA. *Gould.*
PLATE XLVIII.

Z. *corpore toto supra nigrescenti-fusco, plumâ singulâ medio albo-guttatâ et olivaceo-fusco latè marginatâ; remigibus fuscis, mento albo, corpore infra fuscescenti-nigro, gutture pectoreque albo-striatis; abdomine tegminibusque caudæ inferioribus albo irregulariter transversè strigato; rostro obscure corneo; pedibus olivaceo-viridibus.*

Long. tot. 5¾ unc.; *alæ*, 3¼; *caudæ*, 1⅜; *tarsi*, ⅞; *rostri*, ½.

The whole of the upper surface blackish brown, each feather spotted with white down the centre, and largely margined with olive brown; quills plain brown; chin white; the remainder of the under surface brownish black, striated with white on the throat and chest, and crossed by irregular bars of the same on the abdomen and under tail coverts; bill dark horn colour; feet olive green.

Habitat, Rio Plata. (Shot on board the Beagle.)

2. ZAPORNIA SPILONOTA. *Gould.*
PLATE XLIX.

Z. *capite corporeque infra, nigrescenti-griseis; corpore supra obscure rubrofusco, uropygio obscurè grisescenti-nigro; alis hypochondriis postice, tegminibusque caudæ inferioribus albo parciter sparsis; rostro nigrofusco; pedibus rubescentibus; iridibus carmineis.*

Long. tot. 5¼ unc.; *alæ*, 2¾; *caudæ*, 1; *tarsi*, ⅞; *rostri*, ¾.

Head and all the under surface blackish grey; all the upper surface dark reddish brown, fading off on the rump into deep greyish black; the wings, hinder part of the flanks, and under tail coverts slightly sprinkled with white; bill, blackish brown; feet, reddish; iris, bright scarlet.

Habitat, Galapagos Archipelago.

This bird frequents in large numbers the high and damp summits of the islands. It lives in the thick beds of carex and other plants, which, from the condensed vapour of the clouds, are constantly kept rather humid. It is tame, but lives concealed; it often utters a loud and peculiar cry. The female is said to lay from eight to twelve eggs. It is, I believe, the only bird in this archipelago which is exclusively confined to the upper parts of the islands. With respect to the specific description, I must observe, that in one of the specimens, the few and small white spots on the wings and abdomen are wanting. This is not a sexual distinction, but possibly may be owing to immaturity.

1. RALLUS PHILLIPENSIS. *Linn.*

Common on the low coral islets, forming the Keeling or Cocos Atoll in the Indian ocean. With the exception of a snipe, this was the only bird without web-feet which inhabited this group.

2. RALLUS YPECAHA. *Vieill.*

Rallus ypecaha, *Vieill.* Ency. Meth. p. 1071.
Crex melampyga, *Licht.* Cat. Sp.
L'Ypacaha, *Azara*, No. 367.

Buenos Ayres.

3. RALLUS SANGUINOLENTUS. *Swains.*

Rallus sanguinolentus, *Swains*, 2 cent. and a quart.

Valparaiso.

GALLINULA CRASSIROSTRIS. *J. E. Gray.*

Gallinula crassirostris, *J.E. Gray*, in Griff. An. Kingd.

I obtained specimens on the banks of the Plata and at Valparaiso.

FULICA GALEATA. *G. R. Gray.*

Crex galeata, *Licht*, Cat. 80. sp. 826.
Yahana proprement dit, *Azara*, No. 379.
Gallinula galeata, *Bonap.*

Concepcion, Chile.

PORPHYRIO SIMPLEX. *Gould.*

P. vertice, remigibus primariis obscurè olivaceo-viridibus, harum apicibus flavescenti albo anguste marginatis; corpore supra obscure olivaceo-viridi, plumâ singulâ obscurè fulvo late marginatâ; genis gutture, corporeque infra flavescentibus; rostro rubro; pedibus viridescenti-flavis.

Long. tot. 9 unc.; *alæ*, 5¼; *caudæ*, 2½; *tarsi*, 1⅞; *rostrio*, ⅞.

Habitat, Ascension Island, Atlantic Ocean. (*July*.)

This specimen was killed with a stick near the summit of the Island. It was evidently a straggler, which had not long arrived. There is no aboriginal land bird at Ascension.

Order—PALMIPEDES.

Anser melanopterus. *Eyton*.

Anser melanopterus, *Eyton*, Monog. Anatidæ, p. 93.

Plate L.

Captain FitzRoy purchased a skin of this fine goose at Valparaiso, which he has presented to the British Museum. There is another specimen at the Zoological Society, which Mr. Pentland procured from the lake of Titicaca, in Bolivia.

Chloephaga Magellanica. *Eyton*.

Anas Magellanica, *Gmel*. Syst. i. 505.
Chloephaga Magellanica, *Eyton*, Monog. Anatidæ, p. 82.
Bernicla leucoptera, *Less*. Trait d'Ornith. 627.

This goose is found in Tierra del Fuego, and at the Falkland Islands; at the latter it is common. They live in pairs and in small flocks throughout the interior of the island, being rarely or never found on the sea-coast, and seldom even near fresh-water lakes. I believe this bird does not migrate from the Falkland Islands; it builds on the small outlying islets. This latter circumstance is supposed to be owing to the fear of the foxes; and it is perhaps from the same cause, that although very tame by day, they are much the contrary in the dusk of the evening. These geese live entirely on vegetable matter; they are called by the seamen, the "upland geese." Mr. Eyton, in his excellent Monograph on the Anatidæ, has described the trachea of this bird, which I brought home in spirits.

Bernicla antarctica. *Steph*.

Bernicla antarctica, *Steph*. Sh. Zool. xii. 59.
——————*Eyton*, Monograph, p. 84.
Anas Antarctica, *Gmel*. Syst. i. 505.

This goose is common in Tierra del Fuego, the Falkland Islands, and on the western coast, as far north as Chiloe. It is called by the sailors the "rock goose," as it lives exclusively on the rocky parts of the sea-coast. In the deep and retired channels of Tierra del Fuego, the snow-white male, invariably accompanied by his darker consort, and standing close by each other on some distant rocky point, is a common feature in the landscape. Mr. Eyton has described the treachea of this species, which I brought home.

Pæcilonitta Bahamensis. *Eyton.*

Pæcilonitta Bahamensis, *Eyton*, Monog. p. 116.
Anas Bahamensis, *Linn.* Syst. i. 199.
Mareca Bahamensis, *Steph.* Gen. Zool. xii. p. 137.

A specimen was procured from a small salt-water lagoon in the Galapagos Archipelago (*October.*)

It was a male; bill, lead colour; base of superior mandible purple, with a black mark in the upper part.

Dafila urophasianus. *Eyton.*

Dafila urophasianus, *Eyton*, Monog. Anatidæ. p. 112.
Anas urophasianus, *King*, Zool. Journ. iv. 351.

Bahia Blanca, Northern Patagonia.

Rhynchaspis maculatus. *Gould.*

Rhynchaspis maculatus, *Gould*, in Jard. & Selby Illust. Orn. p. 147. pl. 147.

Mr. Gould observes that, "A good figure of this beautiful shoveller may be found in the 3rd vol. of Messrs. Jardine and Selby's Illustrations of Ornithology. Their figure was taken from an example which I forwarded to those gentlemen with the name of *maculata* attached: my specimen was received from the Rio Plata, and this is also the locality whence (in October) Mr. Darwin's specimen was procured. The numerous and conspicuous spots distributed over the body, renders this species readily distinguishable from all the other members of the genus."

1. Querquedula erythrorhyncha. *Eyton.*

Querquedula erythrorhyncha, *Eyton*, Monog. Anatidæ, p. 127.
Anas erythrorhyncha, *Spix*, Av. Nov. sp. pl.

My specimens were obtained from Buenos Ayres (*October*) and the Straits of Magellan (*February.*)

2. Querquedula creccoïdes. *Eyton.*

Querquedula creccoïdes, *Eyton*, Monog. Anatidæ, p. 128.
Anas creccoides, *King*, Zool. Journ. iv. 99.

Mr. Gould observes that, "This species was first described by Mr. Vigors, from a specimen in the collection brought from the Straits of Magellan, by Capt. P. P. King. It is a true teal, and in size and form closely assimilates to the common teal of Europe, and to the species inhabiting North America (*Querquedula Carolinensis*, Bonap.) to both of which it is evidently an analogue, and doubtless represents those birds in the southern half of the American continent." My specimens were procured from the Rio Plata, and from the Straits of Magellan.

MICROPTERUS BRACHYPTERUS. *Eyton.*

Micropterus brachypterus, *Eyton*, Monog. Anat. p. 144.
Anas brachytera, *Lath.* Ind. Orn. ii. 834.

These great logger-headed ducks, which sometimes weigh as much as twenty-two pounds, were called by the old navigators, from their extraordinary manner of paddling and splashing over the water, race-horses, but now much more properly steamers. Their wings are too small and weak to allow of flight, but by their aid, partly swimming and partly flapping the surface of the water, they move very quickly. The manner is something like that by which the common house duck escapes, when pursued by a dog; but I am nearly sure that the steamer moves its wings alternately, instead of, as in other birds, both together. These clumsy birds make such a noise and splashing, that the effect is most curious. The steamer is able to dive but a very short distance. It feeds entirely on shell-fish from the floating kelp and tidal rocks; hence the beak and head are surprisingly heavy and strong, for the purpose of breaking them. So strong is the head, that I have sometimes scarcely been able to fracture it with my geological hammer; and all our sportsmen soon discovered how tenacious these birds were of life. When pluming themselves in the evening in a flock they make an odd mixture of sounds, somewhat like bull-frogs within the tropics.

1. PODICEPS KALIPAREUS. *Quoy & Gaim.*

My specimens were obtained from Bahia Blanca (September), Northern Patagonia, and the Falkland Islands. In the former place it lived in small flocks in the salt-water channels, extending between the great marshes at the head of the harbour. At the Falkland Islands I saw (March) very few individuals; and these only in one small fresh-water lake. Tarsi of the same colour as the plumage of the back; iris of a beautiful tint, between "scarlet and carmine red;" pupil black. Mr. Gould remarks that, "This beautiful species of *Podiceps* is equal in size, and has many of the characters of the *P. auritus*, but is at once distinguished from that species by the silvery colouring of the plumes that adorn the sides of the head; which in *P. auritus* are deep chestnut."

2. PODICEPS ROLLANDII. *Quoy et Gaim.*

Podiceps Rolland, *Quoy et Gaim.* Voy. de l'Uranie, pl. 36. p. 133.

I obtained specimens from the Falkland Islands (March), where it was common at the head of the tortuous bays which intersect those islands; from a fresh water lake near the Strait of Magellan (February); and from the eastern coast of Chiloe. The male and female have the same plumage. Iris of a fine red colour. Mr. Gould adds that, " this species appears to be as nearly related to the *Podiceps cornutus*, as the preceding species is to *P. auritus*, but is readily distinguishable from it, by the white spot in the centre of the tuft of feathers that spring from the sides of the face."

3. PODICEPS CHILENSIS. *Garnot.*

Le macas cornu, *Azara*, No. 443.

This specimen was procured in a fresh-water lake near Buenos Ayres. Capt. P. King brought home specimens from the salt-water channels in Tierra del Fuego, where it is excessively numerous. It often makes a very melancholy cry, which suits the gloomy climate of those desolate shores.

SPHENISCUS HUMBOLDTII. *Meyen.*

Spheniscus Humboldtii, *Meyen.* Nov. Act. Acad. Cæs. Leop. Car. Nat. Cur. 1834, 110. pl. 21.

My specimen was obtained near Valparaiso. Meyen, who first described this bird, procured it from the coast of Peru.

PUFFINUS CINEREUS. *Steph.*

Puffinus cinereus, *Steph.* Gen. Zool. xiii. p. 227.
Procellaria puffinus, *Linn.*

This bird frequents the seas on the whole coast of South America. I obtained specimens from Tierra del Fuego, Chiloe, the mouth of the Plata, and Callao Bay on the coast of Peru. It is likewise known to be common in the Northern Hemisphere; this species, therefore, has a most extensive range. It generally frequents the retired inland sounds in very large flocks; although, occasionally, two or three may be seen out at sea. I do not think I ever saw so many birds of any other sort together, as I once saw of these petrels, behind the Island of Chiloe. Hundreds of thousands flew in an irregular line, for several hours in one direction. When part of the flock settled on the water, the surface was blackened; and a cackling noise proceeded from them, as of human beings talking in the distance. At this time, the water was in parts coloured by clouds of small crustacea. The inhabitants of Chiloe told me that this petrel was very irregular in its movements;—sometimes they appeared in vast numbers, and on the next day not one was to be seen. At Port Famine, every morning and evening, a long band of these birds continued to fly with extreme rapidity, up and down the central parts of the channel, close to the surface of the water. Their flight was direct and vigorous, and they seldom glided with extended wings in graceful curves, like most other members of this family. Occasionally, they settled for a short time on the water; and they thus remained at rest during nearly the whole of the middle of the day. When flying backwards and forwards, at a distance from the shore, they evidently were fishing: but it was rare to see them seize any prey. They are very wary, and seldom approach within gun-shot of a boat or of a ship;—a disposition strikingly different from that of most of the other species. The stomach of one, killed near Port Famine, was distended with seven prawn-like crabs, and a small fish. In another, killed off the Plata, there was the beak of a small cuttle-fish. I observed that these birds, when only slightly winged, were quite incapable of diving. There is no difference in the plumage of the sexes. The web between the inner toes, with the exception of the margin, is "reddish-lilac-purple;" the rest being blackish. Legs and half of the lower mandible blackish purple. From accounts which I have received, the individuals of this species, which live in the Northern Hemisphere, appear to have exactly the same habits as those above described.

1. PELECANOIDES BERARDI. *G. R. Gray.*

Puffinuria Berardi, *Less*. Tr. d'Orn. p. 614.
Procellaria Berardi, *Quoy et Gaim.* Voy. de pl. 31

This bird is common in the deep and quiet creeks and inland seas of Tierra del Fuego, and on the west coast of Patagonia, as far north as the Chonos Archipelago. I never saw but one in the open sea, and that was between Tierra del Fuego and the Falkland Islands. This bird is a complete auk in its habits, although from its structure it must be classed with the Petrels. To the latter Mr. Gould informs me, its affinity is clearly shewn by the form of its beak and nostrils, length of foot, and even by the general colouring of its plumage. To the auks it is related in the general form of its body, its short wings, shape of tail, and absence of hind-toe to the foot. When seen from a distance and undisturbed, it would almost certainly be mistaken, from its manner of swimming and frequent diving, for a grebe. When approached in a boat, it generally dives to a distance, and on coming to the surface, with the same movement takes flight: having flown some way, it drops like a stone on the water, as if struck dead, and instantaneously dives again. No one seeing this bird for the first time, thus diving like a grebe and flying in a straight line by the rapid movement of its short wings like an auk, would be willing to believe that it was a member of the family of petrels;—the greater number of which are eminently pelagic in their habits, do not dive, and whose flight is usually most graceful and continuous. I observed at Port Famine, that these birds, in the evening, sometimes flew in straight lines from one part of the sound to another; but during the day, they scarcely ever, I believe, take wing, if undisturbed. They are not very wild: if they had been so, from their habit of diving and flying, it would have been extremely difficult to have procured a specimen. The legs of this bird are of a "flax-flower blue."

2. Pelecanoides Garnotii. *G. R. Gray.*

Puffinuria Garnotii, *Less.* Voy. de l'Coqu. pl. 46.
Procellaria urinatrix, *Gm.* ?

My specimen was obtained at Iquique (lat. 20° 12′), on the coast of Peru. M. Lesson, who first described this species, says (Manuel d'Ornithologie, vol. ii. p. 394.), " Le *puffinure de Garnot* habite par grandes troupes le long des côtes du Pérou. Il vole médiocrement bien, d'une manière précipitée et en rasant la mer; mais il préfère se tenir en repos sur la surface des eaux, et plonge très fréquemment à la manière des grèbes, sans doute pour saisir les petits poissons qui forment sa pâture." An anatomical description of this bird is there given.

1. Procellaria gigantea. *Gmel.*

This bird, which is called by the English, "Nelly," and by the Spaniards, "Quebranta-huesos," (properly an osprey,) is common in the southern latitudes of South America. It frequents both the inland sounds, and the open ocean far from the coast. It often settles and rests on the water. The Nelly, in its flight and general appearance on the wing, has many points of resemblance with the Albatross; but, as in the case of that bird, it is in vain to attempt observing on what it feeds; both seem to hunt the waters for days together, in sweeping circles, with no success. In the stomach, however, of one which I opened, there was the beak of a large cuttle-fish. The Nelly, moreover, is a bird of prey: it was observed at Port St. Antonio, by some of the officers of the Beagle, to kill a diver. The latter tried to escape, both by diving and flying, but was continually struck down, and at last was killed by a blow on its head. At Port St. Julian, also, these great petrels were seen killing and devouring young gulls. The Nelly breeds on several of the small islands off the coast of Patagonia; for instance, Sea-Lion Island, in the mouth of the Santa Cruz. Most other species of the family retire for the purpose of breeding to the Antarctic Islands.

I have often observed in the southern seas, a bird similar in every respect to the Nelly, excepting in its plumage, being of a much more intense black, and its bill rather whiter. I procured a specimen thus coloured, at Port Famine, and had concluded that it was a distinct species, until Mr. Low, (an excellent practical observer, long acquainted during his sealing voyages with the productions of these seas,) assured me that he positively knew, that these black varieties were the one-year-old birds of the common greyish black Nelly.

2. PROCELLARIA GLACIALOÏDES. *A. Smith.*
Procellaria glacialoïdes, *A. Smith*, Illust. of Zool. of S. Africa, Aves, pl. 51.

I saw this petrel on both sides of the Continent south of lat. 30°; but seldom more than two or three together. I am informed that it arrives in Georgia in September for the purpose of breeding, and that it lays its eggs in holes in the precipices overhanging the sea. On the approach of winter it is said to retire from that island. My specimen was caught in the Bay of St. Mathias (lat. 43° S.) by a line and bent pin, baited with a small piece of pork; the same means by which the Pintado (*Dapt. Capensis*) is so easily caught. It is a tame, sociable, and silent bird; and often settles on the water: when thus resting it might from a distance be mistaken, owing to the general colour of its plumage, for a gull. One or two often approached close to the stern of the Beagle, and mingled with the Pintados, the constant attendants on vessels traversing these southern seas.

DAPTION CAPENSIS. *Steph.*
Procellaria Capensis, *Linn.* Syst. i. 213.

This petrel is extremely numerous over the whole southern ocean, south of the Tropic of Capricorn. On the coast, however, of Peru, I saw them in lat. from 16° to 17° S., which is considerably farther north than they are found on the shores of Brazil. Cook, in sailing south in the meridian of New Zealand, first met this bird in lat. 43° 30'. The Pintados slightly differ in some of their habits from the rest of their congeners, but, perhaps, approach in this respect nearest to *P. glacialoïdes.* They are very tame and sociable, and follow vessels navigating these seas for many days together: when the ship is becalmed, or is moving slowly, they often alight on the surface of the water, and in doing this they expand their tails like a fan. I think they always take their food, when thus swimming. When offal is thrown overboard, they frequently dive to the depth of a foot or two. They are very apt to quarrel over their food, and they then utter many harsh but not loud cries. Their flight is not rapid, but extremely elegant; and as these prettily mottled birds skim the surface of the water in graceful curves, constantly following the vessel as she drives onward in her course, they afford a spectacle which is beheld by every one with interest. Although often spending the whole day on the wing, yet on a fine moonlight night, I have repeatedly seen these birds following the wake of the vessel, with their usual graceful evolutions. I am informed that the Pintado arrives in Georgia for the purpose of breeding, and leaves it, at the same time with the *P. glacialoides*. The sealers do not know any other island in the Antarctic ocean excepting Georgia, where these two birds (as well as the *Thalassidroma oceanica*) resort to breed.

THALASSIDROMA OCEANICA. *Bonap.*

Thalassidroma oceanica, *Bonap.* Journ. Acad. Nat. Scien., Philadelphia, vol. iii. p. 233.
Procellaria oceanica, *Forster.*
Pétrel échasse. *Temm.*

I obtained this bird at Maldonado, near the mouth of the Plata, where it was blown on shore by a gale of wind. These birds, although seeming to prefer on most occasions the open ocean, and to be most active, walking with their wings expanded on the crest of the waves, when the gale is heaviest, yet sometimes visit quiet harbours, in considerable numbers. At Bahia Blanca I saw many, when there was nothing in the weather to explain their appearance. I was informed by a sealer, that they build in holes on the sea cliffs of Georgia, where they arrive very regularly in the month of September. No other place is known to be frequented by them for the purpose of breeding.

PRION VITTATUS. *Cuv.*

Procellaria Vittata, *Gmelin.* Syst. i. 560.

I did not procure a specimen of this bird, although I saw numbers on both sides of the Continent from about lat. 35° S. to Cape Horn. It is a wild solitary bird, appears always to be on the wing: flight extremely rapid. Mr. Stokes (Assistant surveyor of the Beagle) informs me that they build in great numbers on Landfall Island, on the west coast of Tierra del Fuego. Their burrows are about a yard deep: they are excavated on the hill-sides, at a distance even of half a mile from the sea shore. If a person stamps on the ground over their nests, many fly out of the same hole. Mr. Stokes says the eggs are white, elongated, and of the size of those of a pigeon.

1. LARUS FULIGINOSUS. *Gould.*

L. Mas. corpore toto obscurè plumbeo-griseo, tegminibus caudæ superioribus inferioribusque pallidioribus ; rostro basi rubro, apice nigro ; pedibus nigris.

Long. tot. 16½ unc.; alæ, 13½; caudæ, 6; tarsi, 2⅛; rostri 2⅜.

The whole of the plumage deep leaden-grey; the upper and under tail coverts being lightest; bill red at the base, black at the tip; feet black.

Habitat, Galapagos Archipelago (*October*).

This species of gull has many characters in common with the *Larus hæmatorhynchus* of King, from the continent of S. America; but may at once be distinguished from it by the general extreme duskiness of its plumage, feet, tarsi, and bill; and by the more elongated form of the latter. My specimen was killed at James Island. I observed nothing particular in its habits. It is the only species of gull frequenting this Archipelago.

2. LARUS HÆMATORHYNCHUS. *King.*

Larus hæmatorhynchus, *King*, Zool. Journ. iv. 103.
——————————*Jard. & Selb.* Ill. Orn. p. 106.

This bird was killed at Port St. Julian on the coast of Patagonia. Beak (when fresh killed) of a pale "arterial blood red," legs "vermilion red."

3. LARUS DOMINICANUS. *Licht.*

Larus dominicanus, *Licht.* Cat. 82. sp. 846.
Grande Mouette, *Azara*, No. 409.

This gull abounds in flocks on the Pampas, sometimes even as much as fifty and sixty miles inland. Near Buenos Ayres, and at Bahia Blanca, it attends the slaughtering-houses, and feeds, together with the Polybori and Cathartes, on the garbage and offal. The noise which it utters is very like that of the common English gull (*Larus canus*, Linn.).

XEMA (CHROICOCEPHALUS) CIRROCEPHALUM. *G. R. Gray.*

Larus cirrocephalus, *Vieill.* Nov. Dict. d'Histoire, 21. p. 502.
Larus maculipennis, *Licht.* Cat. 83. sp. 855.
Larus glaucodes, *Meyen*, Nov. Act. 1839, p. 115. pl. 24.
Mouette cendrée, *Azara*, No. 410.

This species so closely resembles the *Xema ridibundum*, Boiè, that Mr. Gould observes, he should have hardly ventured to have characterized it as distinct; but as M. Vieillot and Meyen have deemed this necessary, he adopts their view. I have compared a suite of specimens, which I procured from the Rio Plata, the coast of Patagonia, and the Straits of Magellan, with several specimens of the *Xema ridibundum*; the only difference which appears to me constant, is that the primaries of the *X. cirrocephalum*, in the adult winter plumage, both of male and female, are tipped with a white spot (a character common to some other species), whereas in the *X. ridibundum* the points are black. The beak of the latter species, especially the lower mandible, is also a little less strong, or high in proportion to its length. In the immature stage, I could perceive no difference whatever in the plumage of these birds. The proportional quantity of black and white in the primaries, given by Meyen as the essential character, varies in the different states of plumage. The specimens described by this author were procured from Chile.* The soles of the feet of my specimens were coloured, deep "reddish orange," and the bill dull "arterial blood-red" of Werner's nomenclature.

In the plains south of Buenos Ayres I saw some of these birds far inland, and I was told that they bred in the marshes. It is well known that the black-headed gull (*Xema ridibundum*), which we have seen comes so near the *X. cirrocephalum*, frequents the inland marshes to breed. It appears to me a very interesting circumstance thus to find birds of two closely allied species preserving the same peculiarities of habits in Europe and in the wide plains of S. America. Near Buenos Ayres this gull as well as the *L. dominicanus* sometimes attends the slaughter-houses to pick up bits of meat.

RHYNCHOPS NIGRA. *Linn.*

I saw this bird both on the East and West coast of South America, between latitudes 30° and 45°. It frequents either fresh or salt water. Near Maldonado (in May), on the borders of a lake, which had been nearly drained, and which in consequence swarmed with small fry, I watched many of these birds flying backwards and forwards for hours together, close to its surface. They kept their bills wide open, and with the lower mandible half buried in the water. Thus skimming the surface, generally in small flocks, they ploughed it in their course; the water was quite smooth, and it formed a most curious spectacle, to behold a flock, each bird leaving its narrow wake on the mirror-like surface. In their flight they often twisted about with extreme rapidity, and so dexterously managed, that they ploughed up small fish with their projecting lower mandibles, and secured them with the upper half of their scissor-like bills. This fact I repeatedly witnessed, as, like swallows, they continued to fly backwards and forwards, close before me. Occasionally, when leaving the surface of the water, their flight was wild, irregular, and rapid; they then also uttered loud harsh cries. When these birds were seen fishing, it was obvious that the length of the primary feathers was quite necessary in order to keep their wings dry. When thus employed, their forms resembled the symbol, by which many artists represent marine birds. The tail is much used in steering their irregular course.

These birds are common far inland, along the course of the Rio Parana; and

---

* The naturalists in Lutke's voyage, vol. iii. p. 255, seem to consider a gull, which they obtained at Concepcion, as the *Larus Franklinii* of North America.

it is said they remain there during the whole year, and that they breed in the marshes. During the day they rest in flocks on the grassy plains, at some distance from the water. Being at anchor in a small vessel, in one of the deep creeks between the islands in the Parana, as the evening drew to a close, one of these scissor-beaks suddenly appeared. The water was quite still, and many little fish were rising. The bird continued for a long time to skim the surface; flying in its wild and irregular manner up and down the narrow canal, now dark with the growing night and the shadows of the overhanging trees. At Monte Video, I observed that large flocks remained during the day on the mud banks, at the head of the harbour; in the same manner as those which I observed on the grassy plains near the Parana. Every evening they took flight in a straight line seaward. From these facts, I suspect, that the Rhynchops frequently fishes by night, at which time, many of the lower animals come more abundantly to the surface than during the day. I was led by these facts to speculate on the possibility of the bill of the Rhynchops, which is so pliable, being a delicate organ of touch. But Mr. Owen, who was kind enough to examine the head of one, which I brought home in spirits, writes to me, (August 7, 1837,) that—

"The result of the dissection of the head of the *Rhynchops*, comparatively with that of the head of the duck, is not what you anticipated. The facial, or sensitive branches of the fifth pair of nerves, are very small; the third division in particular, is filamentary, and I have not been able to trace it beyond the soft integument at the angles of the mouth. After removing with care, the thin horny covering of the beak, I cannot perceive any trace of those nervous expansions which are so remarkable in the lamelli-rostral aquatic birds; and which in them supply the tooth-like process, and soft marginal covering of the mandibles. Nevertheless, when we remember how sensitive a hair is, through the nerve situated at its base, though without any in its substance, it would not be safe to deny altogether, a sensitive faculty in the beak of the Rhynchops."

M. Lesson (Manuel d'Ornithologie, vol. ii. p. 385.) has stated, that he has seen these birds opening the shells of the Mactræ, buried in the sandbanks on the coast of Chile. From their weak bills, with the lower mandible so much produced, their short legs and long wings, it seems very improbable that this can be a general habit, although it may sometimes be resorted to. Wilson, who was well acquainted with this bird, does not believe "the report of its frequenting oyster beds, and feeding on these fish." The existence, however, of this same report in the United States, makes the question, whether the Rhynchops does not sometimes turn the peculiar structure of its beak to this purpose, worthy of further investigation.

VIRALVA ARANEA. *G. R. Gray.*

Sterna aranea, *Wils.* Am. Orn. pl. 72. f. 6.

My specimen was procured at Bahia Blanca, in Northern Patagonia. I may here observe, that many navigators have supposed that terns, when met with out at sea, are a sure indication of land. But these birds seem not unfrequently to be lost in the open ocean; thus one (*Megalopterus stolidus*) flew on board the Beagle in the Pacific, when several hundred miles from the Galapagos Archipelago. No doubt, the remark made by navigators, with respect to the proximity of land where terns are seen, refers to birds in a flock, fishing, or otherwise showing that they are familiar with that part of the sea. I, therefore, more particularly mention, that off the mouth of the Rio Negro, on the Patagonian shore, I saw a flock (probably the *Viralva aranea*) fishing seventy miles from land: and off the coast of Brazil a flock of another species, 120 from the nearest part of the coast. The latter birds were in numbers, and were busily engaged in dashing at their prey.

MEGALOPTERUS STOLIDUS. *Boiè.*

Sterna stolida, *Linn.* Syst. i. 227.

My specimens were procured from the Galapagos Archipelago. It is well known to be an inhabitant of the seas in the warmer latitudes over the whole world. The Rocks of St. Paul's, nearly under the equator, in the Atlantic ocean, were almost covered with the rude and simple nests of this bird, made with a few pieces of sea-weed. The females were sitting upon their eggs (in February), and by the side of many of their nests, parts of flying-fish were placed, I suppose, by the male bird for his partner to feed on during the labour of incubation.

PHALACROCORAX CARUNCULATUS. *Stephens.*

Phalacrocorax carunculatus, *Steph.* Gen. Zool.
Pelecanus carunculatus, *Gm.* Syst. i. 576.
Phalacrocorax imperialis, *King,* Zool. Proc. vol. i. pt. 1. 30.

I procured a specimen of this bird at Port St. Julian, on the coast of Patagonia, where, during January, many were building. I merely mention it here, for the purpose of describing the singularly bright colours of the naked skin about its head. Skin round the eyes "campanula blue;" cockles at the base of the upper mandible, "saffron mixed with gamboge-yellow." Marks between the eye and the corner of the mouth, "orpiment orange;" tarsi scarlet.

FREGATA AQUILA. *Cuv.*

Pelecanus Aquilus, *Linn.*

I had an opportunity, at the Galapagos Archipelago, of watching, on several occasions, the habits of this bird, which are very interesting in relation to its peculiar structure. The Frigate bird, when it sees any object on the surface of the water, descends from a great height, in an inclined plane, head foremost, with the swiftness of an arrow; and at the instant of seizing with its long beak and outstretched neck, the floating morsel, it turns upwards, with extraordinary dexterity, by the aid of its forked tail, and long, powerful wings. It never touches the water with its wings, or even with its feet; indeed I have never heard of one having been seen on the surface of the sea; and it appears that the deeply indented web between its toes is of no more use to it, than are the shrivelled wings beneath the wing-cases of some coleopterous beetles. The Frigate bird has a noble appearance when seen soaring in a flock at a stupendous height (at which time it merits the name of the Condor of the ocean), or when many together are dashing, in complicated evolutions, but with the most admirable skill, at the same floating object. They seem to scorn to take their food quietly, for between each descent they raise themselves on high, and descend again with a swift and true aim. If the object (such as offal thrown overboard) sink more than six or eight inches beneath the surface, it is lost to the Frigate bird. I was informed at Ascension, that when the little turtles break through their shells, and run to the water's edge, these birds attend in numbers, and pick up the little animals (being thus very injurious to the turtle fishery) off the sand, in the same manner as they would from the sea.

APPENDIX.

Anatomical description of *Serpophaga albocoronata, Furnarius cunicularius, Uppucerthia dumetoria, Opetiorhynchus vulgaris, O. antarcticus, O. Patagonicus, Pteroptochos Tarnii, P. albicollis, Synallaxis maluroides, Phytotoma rara, Trochilus gigas, Tinochorus rumicivorus.**

BY T. C. EYTON, Esq., F.L.S., &c.

SERPOPHAGA ALBOCORONATA. *Gould.* (Male.)

Tongue pointed, furnished with a few short bristles at the sides near the base. Trachea with the same muscles as among the warblers generally. Æsophagus slightly funnel-shaped; proventriculus much expanded at its entrance into the gizzard, which is rounded, not very muscular, inner coat slightly hardened, smooth. Intestine of moderate size, furnished with two rudimentary cæca.

	inches		inches
Length of œsophagus, including proventriculus	1	Length of intestine from gizzard to cloaca	$3\frac{1}{4}$
of gizzard	$\frac{3}{8}$	from cæca to cloaca	$3\frac{1}{4}$
Breadth of ditto	$\frac{7}{16}$		

The skeleton of this bird is precisely that of the smaller and weaker species of Laniadæ.

	lines		
Length of sternum	5	No of cervical vertebræ	11
Breadth anteriorly	3	dorsal ditto	7
posteriorly	$4\frac{1}{4}$	sacral ditto	9
Width of fissures	1	caudal ditto	6
Depth of ditto	$1\frac{1}{2}$		Total 33
Depth of keel	2		
Length of pelvis	$5\frac{1}{4}$		
Width anteriorly	$2\frac{1}{4}$	No. of false ribs	1 1?
posteriorly	$5\frac{1}{4}$	true ditto	5
Length from occiput to point of bill	12		Total 7
Breadth of head	$5\frac{3}{4}$		
Length of coracoids	$4\frac{1}{4}$		

* I am much indebted to Mr. Eyton for these observations, which greatly add to the value of the previous descriptions

FURNARIUS CUNICULARIUS. *G. R. Gray.* (Male.)

Tongue, trachea, and œsophagus, as in *Uppucerthia.* Proventriculus longer, and slightly contracted at its entrance into the gizzard, which is large, flattened, and muscular, more rounded than in *Opetiorhynchus,* lined with a rugose hardened coat, and filled with small seeds, and the remains of insects ; intestines of small diameter, and furnished with two rudimentary cæca.

	inches		inches
Length of œsophagus, including proventriculus	1¾	Length from gizzard to cæca	5
of gizzard	¾	cæca to cloaca	1¼
Breadth of ditto	⅝		

Sternum of nearly equal breadth, both posteriorly and anteriorly, but much narrowed in the middle, the portion to which the ribs are attached much elongated beyond their junction ; posterior margin furnished with two deep fissures, slightly narrowed at their exit ; keel deep, slightly rounded on its inferior edge, and much scolloped out anteriorly ; pelvis broad and short, the os pubis projecting far backwards ; the ischium terminating posteriorly in an acute process.

Os furcatum thin, much arched, furnished with a flattened reflexed process at its junction with the sternum ; the points of the rami bent forwards at their junction with the coracoids.

Coracoids of moderate size and length, inserted deeply into the sternum ; scapula of moderate size, broader near the extremity.

	lines		
Length of sternum	11	No. of cervical vertebræ	12
Breadth anteriorly	6¼	dorsal ditto	7
posteriorly	8¼	sacral ditto	10
Depth of keel	4½	caudal ditto	7
Length of pelvis	12		
Width anteriorly	4½	Total	36
posteriorly	11		
Length from occiput to point of bill	10	No. of true ribs	5
Breadth of cranium	7½	false ditto	2·1
Length of coracoids	8	Total	8

UPPUCERTHIA DUMETORIA. *Geoff. & D'Orb.* (Female.)

Tongue short, compared with the length of the bill, pointed, armed with a few spines at the base ; trachea of moderate size, acted upon by one pair of sterno-tracheal muscles, which go off to the sternum, about ⅛ of an inch above the inferior larynx ; from the upper ring of the bronchiæ on each side, a process proceeds upwards to the point from which the muscles diverge, to which point only the rings of the trachea are continued, two spaces therefore, one on the anterior, the other on the posterior side of the trachea, immediately above the bronchiæ, are left devoid of osseous matter, being bounded laterally by the process above mentioned, inferiorly by the upper rings of the bronchiæ, and superiorly by the lower ring of the trachea, which is slightly enlarged ; œsophagus small, slightly dilated a little above the proventriculus, which is of moderate size, and not contracted before entering the gizzard ; gizzard large, oval, very muscular, inner coat hardened, deeply furrowed longitudinally, and filled with the remains of insects ; intestinal canal of moderate size, without cæca ; rectum very slightly enlarged ; liver bilobed.

	inches		inches
Length of œsophagus, including proventiculus	2	Breadth of ditto	½
of gizzard	¾	Length of intestinal canal	10

With the exception of being larger than *Furnarius cunicularius*, and in having the bill more bent and longer, the skeleton presents no material difference from that of the above-named bird.

	lines		
Length of sternum	13	No. of cervical vertebræ	11
Breadth anteriorly	6	dorsal ditto	7
posteriorly	7¼	sacral ditto	11
Depth of keel	4	caudal ditto	6
of fissures	4	Total	35
Breadth of ditto	1		
Length of pelvis	14¼	No. of true ribs	5
Breadth anteriorly	4	false ditto	2·1
posteriorly	9¼	Total	8
Length from occiput to point of bill	27		
Breadth of cranium	8		
Length of coracoids	11		

OPETIORHYNCHUS VULGARIS, *Gray*. (Male.)

The structure of the soft parts, both in this species of *Opetiorhynchus*, and the two following ones, so closely resemble that of *Furnarius* and *Uppucerthia*, that one description will almost serve for the whole; those differences that do exist being not more than are generally found in species of the same genus; the external characters also being slight, I cannot but doubt the propriety of separating them; the cæca are slightly developed in this species, measuring ⅛ inch in length.

	inches		inches
Length of œsophagus, proventriculus included	2¼	Length of intestinal canal from gizzard to the cloaca	7½
of gizzard	⅝	from cæca to cloaca	⅞
Breadth of ditto	¼		

Skeleton similar in form to that of *Furnarius cunicularius*.

	lines		
Length of sternum	11¾	No. of cervical vertebræ	11
Breadth anteriorly	5¼	dorsal ditto	7
posteriorly	7½	sacral ditto	11
Depth of keel	3¾	caudal ditto	7
of fissures	5	Total	36
Breadth of ditto	1½		
Length of pelvis	12¾	No. of true ribs	5
Breadth anteriorly	4	false ditto	2·1
posteriorly	9¾	Total	8
Length from occiput to point of bill	17		
Breadth of cranium	7		
Length of coracoids	8½		

OPETIORHYNCHUS ANTARCTICUS. *G. R. Gray*. (Male.)

Structure of the soft parts as in *O. vulgaris*, but with the rectum of rather larger diameter, and the cæca very minute; gizzard filled with the remains of insects.

	inches		inches
Length of œsophagus, including proventriculus	2¼	Breadth of gizzard	½
gizzard	⅝	Length of intestinal canal from gizzard to cloaca	7

Skeleton similar in form to *Furnarius cunicularius*, and the other species of this genus.

	lines		
Length of sternum	11	No. of cervical vertebræ	11
Breadth anteriorly	6	dorsal ditto	7
posteriorly	7½	sacral ditto	12
Depth of keel	4¾	caudal ditto	7
of fissures	4	Total	37
Breadth of ditto	1¾		
Length of pelvis	12	No. of true ribs	5
Breadth anteriorly	3¾	false ditto	2·1
posteriorly	10½	Total	8
Length from occiput to point of bill	18		
Breadth of cranium	7½		
Length of coracoids	9		

OPETIORHYNCHUS PATAGONICUS. *G. R. Gray.* (Male.)

No difference in the structure of the soft parts from the other species of the genus before spoken of. The trachea, however, does not differ from the ordinary simple form found in most birds, but differs from *O. vulgaris* and *O. antarcticus*, in having the lower rings continued to the bronchiæ it is acted upon by one pair of muscles; no cæca are apparent.

	inches		inches
Length of œsophagus, including proventriculus	2⅛	Breadth of gizzard	⅜
gizzard	½	Length of cutis from gizzard to cloaca	5½

Skeleton in form similar to that of *Furnarius cunicularius*, and the other species of this genus.

	lines		
Length of sternum	13	No. of cervical vertebræ	11
Breadth anteriorly	6½	dorsal ditto	7
posteriorly	8½	sacral ditto	9
Depth of keel	5	caudal ditto	6
fissures	4	Total	33
Breadth of ditto	1½		
Length of pelvis	13½		
Breadth anteriorly	5	No. of true ribs	5
posteriorly	10½	false ditto	2·1
Length from occiput to point of bill	19	Total	8
Breadth of cranium	8		
Length of coracoids	10		

Remarks:—the last five species approach so nearly, that I doubt the propriety of separating them generically. The skeletons are only distinguishable with the exception of the form of the bill, by the proportions between the different admeasurements.

PTEROPTOCHOS TARNII. *G. R. Gray.* (Female.)

Tongue pointed, armed with two strong lateral spines, and a few intermediate smaller ones at the base; œsophagus largest at the upper extremity, and gradually becoming smaller towards the proventriculus; no vestige of a craw; proventriculus of moderate size, not much contracted towards the gizzard, which is also of moderate size, and much flattened; not very muscular, and lined with a hardened coat, rugose longitudinally; the gizzard was filled with small pebbles, and a coarse black powder, probably the remains of insects; intestinal canal small; cæca rudimental; rectum large, becoming more expanded towards the cloaca, which is also large; trachea of equal diameter throughout, furnished with one pair of sterno-tracheal muscles, a portion of each of which is continued downwards to the upper rings of the bronchiæ, on which it expands; liver two-lobed.

	inches		inches
Length of œsophagus, including proventriculus	3½	Diameter of gizzard	9/16
of intestinal canal, from gizzard to cloaca	18	Length of ditto	1
of rectum	2¼		

The pelvis and ribs of this bird were much damaged; sternum of equal breadth posteriorly and anteriorly, slightly contracted on its lateral edge, near the middle indented on its posterior margin with four deep fissures, the outer ones largest; a large triangular process projecting forwards between the junctions of the coracoids, bifid at the apex; the coracoids themselves very strongly articulated to the sternum, the sides of the sternum to which the ribs are articulated projecting in the form of a process far beyond the junction of the coracoids; the sternal keel is narrow, and has its edge straight; the coracoids are long, thin, with very slight external lateral processes at their junction with the sternum; os furcatum very thin, roundish, a very slight process on the point at which it approaches nearest to the sternum, very slightly arched.

Scapula broad, flattened, much widened at about one-third of its length from the hinder extremity; wing bones short, and weak; leg bones long, and strong; the fibula much developed.

	lines		lines
Length of sternum	15	Length from occiput to point of bill	22¼
Greatest breadth of sternum	9½	Breadth of cranium	10½
Breadth at the narrowest part	7	Length of coracoids	11
Width of external fissure	1½	Breadth of scapula in the broadest part	2
Depth of ditto	6	Cervical vertebræ	12
Width of internal ditto	1¼	Dorsal ditto	6
Depth of ditto	6½	Sacral, damaged.	
Depth of keel	3	Caudal, damaged.	

PTEROTOCHOS ALBICOLLIS. *Kittl.* (Male.)

Trachea, tongue, œsophagus, gizzard, and liver of the same form as in *Pterotochos Tarnii.* The contents of the gizzard also did not differ.

	inches		inches
Length of intestinal canal	14½	Length of gizzard	¾
from cæca to cloaca	2½	Breadth of ditto	⅝

Only the body, after skinning, of the species, was brought home by Mr. Darwin.

The skeleton of this species does not differ in anything but admeasurements from that of *Pterotochos Tarnii;* the pelvis, however, being so much damaged in that species, that I was not able to make many notes upon it, I shall give a description of that part in the present one.

Pelvis of moderate size; the ossa pubis and ischium much expanded, and elongated posteriorly, and placed nearly perpendicular to the plane of the ilium, ischiatic foramina large; two large processes arise on the ilium on each side of the junction of the caudal vertebræ for the attachment of the levator muscles of the tail.

APPENDIX.

	lines
Length of pelvis	14
Breadth posteriorly	8½
anteriorly	4
Length of sternum	9½
Breadth of ditto	7
in the narrowest part	5½
Depth of keel	2½
Length of coracoids	7½
Breadth of scapula in the widest part	1

No. of cervical vertebræ, wanting.	
dorsal ditto, wanting.	
sacral ditto	9
caudal ditto, wanting.	

Remarks:—Both this and the foregoing bird are most curious; it is difficult to say in what order they ought to be placed, the external form being equally ambiguous with the internal structure.

The digestive organs nearly agree with that of many insessorial birds; the pelvis also approaches nearly in form to that of the thrush; the sternum, however, differs altogether from any form found in that order, and is precisely that of a *Picus*. The coracoids are lengthened; the os furcatum is furnished with only a slight process where it approaches the sternum, in which particulars, also in the form of the ribs, it agrees with the *Picidæ*.

Having found the internal structure so curious, and so contrary to what I expected, I was led to examine the external more minutely than I had before done. The same form of claw is found in several species among the cuckoos, in the genus *Pelophilus*, for instance; the two outer toes are zygodactyle, being united together as far as the first joint; the bill, at first sight, might be taken for that of a gallinaceous bird; but in searching among the order *Scansores*, for some resemblance, I find the same structure in several of the cuckoo family, with the exception of the nostrils being nearer to the apex of the bill in *Pterotochos*. The Australian genus *Menura* is, probably, allied to this, but differs in the structure of the nostrils.*

SYNALLAXIS MALUROIDES. *D'Orb.* (Female.)

Tongue pointed, furnished at the base with two strong spines, the sides of which are armed with smaller ones; trachea, œsophagus, and proventriculus as in *Furnarius* and *Uppucerthia*; gizzard rounded, not very muscular, lined with a slightly hardened smooth coat, and filled with the remains of insects; intestinal canal of moderate size and length, furnished with two rudimentary cæca.

	inches			inches
Length of œsophagus and proventriculus	1¼		Length of intestinal canal from gizzard to cloaca	4¾
gizzard	⅜		from cæca to cloaca	¾
Breadth of ditto	1/12			

The parts of the skeleton of this bird which I was able to preserve, were more closely allied to the corresponding ones of Troglodytes than to those of any other genus in my possession, but differ in the following particulars: the lateral processes of the sternum bounding the posterior fissures are not so much expanded, consequently the fissures themselves are smaller; the keel is rather deeper; the portion to which the ribs are attached does not project so far forwards, but the

---

* Since the above was in type, I have had, through the kindness of Mr. Gould, an opportunity of examining *Menura lyra*, and find my former supposition to be correct; but neither of these genera can be placed among the gallinaceous birds where the latter bird has been arranged by some authors.

process between the coracoids is rather longer; the interocular portion of the cranium is also rather broader than in the above-mentioned genus; the pelvis, coracoids, and scapula agree both in shape and size with Troglodytes.

	lines		lines
Lenth of sternum	6¼	Breadth of cranium	5¾
Breadth anteriorly	4	Length of pelvis	9
posteriorly	4½	Breadth of ditto posteriorly	5
Greatest width of fissures	⅞	anteriorly	1¾
Depth of ditto	2¼		
Length of occiput to point of bill	14¾	No. of cervical vertebræ	12

PHYTOTOMA RARA. *Molina.*

This bird being injured about the sexual organs, I could not ascertain the sex. Tongue pointed, armed at the base on each side with a flattened tricuspid spine; trachea small, of uniform diameter throughout its whole length, acted upon by one pair of sterno-tracheal muscles; œsophagus funnel-shaped at the upper extremity, when distended capable of containing a common pencil, at its junction with the proventriculus much smaller; proventriculus scarcely perceptible; gizzard small, consisting of little more than a thick skin, inner coat hardened; the entrance of the œsophagus, and the exit of the intestine placed very near together at the upper extremity of it; intestinal canal very short, and of large diameter, entirely devoid of cæca; the whole length with the gizzard and œsophagus distended with a stringy substance, resembling coarse spun cotton cut into short lengths.

	inches		inches
Length of œsophagus, including proventriculus	3	Length of gizzard	⅞
of intestinal canal	7½	Breadth of ditto	½

Sternum of nearly equal breadth, both posteriorly and anteriorly, much narrowed near the middle; posterior margin nearly straight, indented with two large fissures, narrowed at their exit; between the junctions of the coracoids furnished with a bifid process; the portion of the sternum to which the ribs are attached, continued anteriorly beyond the junction of the coracoids; keel of moderate size; coracoids long, not very strong; os furcatum long, slightly arched, furnished with a flattened process, turned inwards at the point it approaches the sternum.

Pelvis broad, and short, narrowest anteriorly, the os pubis and ischium continued far backwards, beyond the junction of the caudal vertebræ; ribs strong, and flattened; posterior process large; scapula long, broadest near the extremity; legs of moderate strength, the internal processes of the tibia large, and flattened; bones of the cranium strong.

	lines		
Length of sternum	12½	No. of cervical vertebræ	11
Breadth anteriorly	6¼	dorsal ditto	7
posteriorly	9¼	sacral ditto	10
Width of fissures	1¼	caudal ditto	7
Depth of ditto	4		
keel	4¾	Total	35
Length of pelvis	13¼		
Width anteriorly	5	No. of true ribs	5
posteriorly	11	false ribs	2·1
Length from occiput to point of bill	16		
Breadth of head	8	Total	8
Length of coracoids	9		

Remarks:—The skeleton and soft parts of this bird very nearly resemble those of the genus *Loxia*, but differ in their superior size, in having the fissures on the posterior margin of the sternum not so deep, and in the margin itself being straighter, the coracoids larger, and in having the process at the end of the os furcatum approaching the sternum smaller than in that genus. The ribs also are stronger.

TROCHILUS GIGAS. *Vieill.* (Male.)

Tongue bifid, each division pointed; hyoids very long, in their position resembling those in the *Picidæ;* trachea of uniform diameter; destitute of muscles of voice; bronchia very long; œsophagus funnel-shaped, slightly contracted on approaching the proventriculus, which is small, and scarcely perceptible; gizzard small, moderately muscular, the inner coat slightly hardened, and filled with the remains of insects; intestine largest near the gizzard; I could not perceive a vestige of cæca.

	inches		inches
Length of œsophagus, including proventriculus	1¾	Length of gizzard	¼
intestinal canal	3½	Breadth of ditto	⅓

Sternum with the keel very deep, its edge rounded, and projecting anteriorly; posterior margin rounded, and destitute of indentation or fissure; the ridges to which the pectoral muscles have their attachment, large and prominent, the horizontal portion much narrowed anteriorly, consequently the junctions of the coracoids are very near together.

Pelvis short, very broad; os pubis long, curved upwards at the extremities, projecting far downwards, and posteriorly beyond the termination of the caudal vertebræ; the ischiatic foramen small, and linear; femora placed far backwards; coracoids short, very strong, their extremities much diverging; os furcatum short, slightly arched near the extremities of the rami, which are far apart, furnished with only a small process on its approach to the sternum; scapula flattened, long, broadest near the extremity; humerus, radius, and ulna short, the metacarpal bones longer than either; the former furnished with ridges much elevated for the attachment of the pectoral muscles; caudal and dorsal vertebræ with the transverse processes long, and expanded; cranium of moderate strength, the occipital portion indented with two furrows, which pass over the vertex, and in which the hyoids lie; orbits large, divided by a complete bony septum; the lachrymal bones large, causing an expansion of the bill near the nostrils.

	lines		
Length of sternum	13¼	No. of cervical vertebræ	10
Breadth anteriorly	4	dorsal ditto	6
posteriorly	7¼	sacral ditto	9
Depth of keel	6¾	caudal ditto	5
Length of pelvis	6¼	Total	30
Width anteriorly	2½		
posteriorly	7		
Length from occiput to point of bill	27½	No. of true ribs	5
Breadth of cranium	6⅓	false ditto	1·3
Length of coracoids	6	Total	9

Remarks:—The skeleton of this bird does not differ in form from that of *Trochilus pella*, figured at page 270 of the Cyclopædia of Anatomy and Physiology. The whole of the group are more nearly allied to fissirostral birds than any other.

TINOCHORUS RUMICIVORUS. *Eschsch.* (Male.)

Trachea of uniform diameter, furnished with one pair of sterno-tracheal muscles, from which a few fibres descend on each side to the upper rings of the bronchiæ; œsophagus of large diameter to about half its length, where it is furnished with a craw, and afterwards contracted to the proventriculus; the craw where it is connected with the œsophagus is much contracted, afterwards it expands into a large sac; proventriculus small; gizzard large, and very muscular; the grinding surfaces hard, concave in the middle, and furnished with longitudinal grooves in the concave part; the intestinal canal is of moderate length, small next the gizzard, largest at the entrance of the cæca, from whence it slightly tapers to the cloaca, which is small; cæca long, of greatest diameter at the opposite extremity to their entrance into the rectum; the gizzard and œsophagus were filled with reeds, mixed with very small pebbles; liver bilobed.

Length of œsophagus from glottis to gizzard	3 inch.	Diameter parallel to the grinding surfaces	¾ inches
from œsophagus to outer extremity of craw	¾ inch.	Length of intestine from gizzard to cloaca	13
Perpendicular diameter of craw	7 lines	from cæca to cloaca	1¾
Greatest diameter of gizzard obliquely to the grinding surfaces	1 inch.	of cæca	3

A second specimen, a female, did not differ, except in sex. Skeleton light; bones in general thin. Sternum broadest posteriorly, and indented on its posterior margin with two large fissures; keel deep, its inferior edge rounded, much scolloped out anteriorly; a moderate-size bifid manubrial process between the junction of the coracoids.

Pelvis broad, of moderate length, similar to that found among the genus *Strepsilas*.

Os furcatum much arched, furnished with a small flattened process, where the ligament unites it to the sternum; coracoid of moderate length, strong, furnished with a large process externally near their junction with the sternum; ribs flattened, posterior process long, slightly curved, and narrow.

	lines		
Length of sternum	16	No. of cervical vertebræ	14
Breadth anteriorly	7	dorsal	6
posteriorly	11	sacral	12
Width of fissures	4	caudal	7
Depth of ditto	6	Total	39
keel	7		
Length of pelvis	10½		
Width anteriorly	6	No. of true ribs	6
posteriorly	12	false ditto	2
Length from occiput to point of bill	16	Total	8
Breadth of head	6¼		
Length of coracoids	7¼		

Remarks. The bill of this curious bird much resembles that of the genus *Glareola*, but the soft skin covering the nostrils is more developed, in which respect it resembles the quails, and other gallinaceous birds. The structure of the tarsi, feet, and nails approach near to that of *Strepsilas*, but differ in the latter being sharper, and in the scales on the feet and tarsi being more apparent, which may, perhaps, have been caused to a certain degree by the bird having been for a long while in spirits.

The wing has precisely the same structure as in *Glareola*, and some of the plovers.

The tail is more lengthened than among the plovers, but not more so than in *Glareola praticola*, which species has, however, the tail forked, but some of the same genus, as the last named bird, although it is not so long in them, have it in the same shape as in *Tinochorus*,— as *Glarecola Australis*.

The structure of the digestive organs is altogether that of a gallinaceous bird; the skeleton, however, agrees scarcely in any particular with that order, approaching closely to that of the waders. The sternum differs from any gallinaceous bird with which I am acquainted, in wanting entirely the strong lateral process, and in the fissures on the posterior margin being much smaller; the nearest approach in form which I have been able to find, is that of *Machetes*, from which, if it were not for the superior size of the latter, it could scarcely be distinguished.

The pelvis agrees so perfectly with that of *Strepsilas interpres*, and the *Charadriidæ* in general, as not to require farther remark.

The remainder of the skeleton resembles both the plovers and sandpipers.

I much regret that I have never had an opportunity of dissecting a specimen of *Glareola*, to which the genus, *Tinochorus*, appears closely allied, and I believe that they will form a connecting link between the orders *Grallatores* and *Razores*.

INDEX TO THE SPECIES.

N.B. The Synonyms are in Italics.

	Page		Page
Ada Commersoni	51	Anas Magellanica	134
Agelaius chopi	107	—— urophasianus	135
—— fringillarius	106	—— Bahamensis	135
—— chrysopterus	106	Anser melanopterus	134
—— virescens	107	Anthus furcatus	85
Aglaia striata	97	—— Chii	85
—— vittata	98	—— fulvus	84
Agriornis	55	—— variegatus	84
—— gutturalis	56	—— correndera	85
—— striatus	56	Anumbi rouge	80
—— micropterus	57	Anumbius ruber	80
—— maritimus	57	Aquila pezopora	13
Agriornis leucurus	57	—— megaloptera	21
Alaudinæ	87	Ardea leuce	128
Alauda cunicularia	65	—— Egretta	128
—— nigra	84	—— Herodias	128
—— rufa	84	—— nycticorax	128
—— fulva	84	—— violacea	128
Albatuitui à collier noir	127	—— callocephala	128
Alcedo torquata	42	Athene cunicularia	31
—— Americana	42	Attagis Gayii	117
—— Senegalensis	41	—— Falklandica	117
Alecturus guirayetupa	51	Becassine, 1ᵉ Espece	131
Alouette noire à dos fauve	84	Bec d'argent	51
—— à dos rouge	84	Bernicla leucoptera	134
Amblyramphus bicolor	109	—— antarctica	134
—— ruber	109	Buteo tricolor	26
Ammodramus Manimbè	90	—— varius	26
—— xanthornus	90	—— ventralis	26
—— longicaudatus	90	—— erythronotus	26
Anas Antarctica	134	Buteoninæ	22
—— brachyptera	136	Cactornis assimilis	105
—— erythrorhyncha	135	—— scandens	104
—— creccoides	135	Caille des Isles Malouines	117

INDEX TO THE SPECIES.

	Page		Page
Camarhynchus psittaculus	103	Columba gymnopthalmus	115
——— crassirostris	103	——— leucoptera	115
Caprimulgidæ	36	——— picazuro	115
Caprimulgus bifasciatus	36	——— Talpacoti	116
——— parvulus	37	——— Boliviana	116
Cathartes aura	8	——— aurita	115
——— atratus	7	Columbina strepitans	116
——— urubu	7	——— talpacoti	116
Certhia antarctica	67	——— Cabocolo	116
Certhidea olivacea	106	Condor	3
Certhilauda cunicularia	65	Conurus murinus	112
Ceryle torquata	42	——— Patachonicus	113
——— Americana	42	La Correndera	85
Charadrius virginius	126	Coturnix Falklandica	117
——— marmoratus	126	Craxirex Galapagoensis	23
——— semipalmatus	128	Crex lateralis	132
——— bifasciatus	127	——— galeata	133
——— trifasciatus	127	——— melampyga	133
——— rubecola	126	Crithagra ? brevirostris	88
——— Azaræ	127	——— Braziliensis	88
——— collaris	127	Crotophaga ani	114
——— Cayanus	127	——— Piririqua	114
Charpentier des champs	113	Cryptura Guaza	120
Chii	85	Crypturus rufescens	120
Chingolo	91	——— perdicarius	119
Chionis alba	118	Cuculus guira	114
Chloephaga Magellanica	134	——— nævius	114
Chlorospiza xanthogramma	96	Culicivora parulus	49
——— melanodera	95	Curruca macloviana	83
Chopi	107	Cursores	120
Chrysometris campestris	89	Cyanotis omnicolor	86
——— Magellanica	97	Cyclarhis Guianensis	58
Chrysoptilus campestris	113	Cypselus unicolor	41
Churrinche	44	Dafila urophasianus	135
Circaëtus antarcticus	15	Daption Capensis	140
Circinæ	29	Dendrodramus	82
Circus histrionicus	50	——— leucosternus	82
——— megaspilus	29	Diplopterus nævius	114
——— cinereus	30	——— guira	114
Clignot ou Lichenops	51	Dolichonyx oryzivorus	106
Coccothraustinæ	98	Dragon	107
Colaptes Chilensis	114	Egretta leuce	128
Columba Fitzroyii	114	Emberiza melanodera	95
——— denisea	114	——— carbonaria	94
——— araucana	114	——— guttata	94
——— loricata	115	——— Manimbé	90

INDEX TO THE SPECIES.

	Page		Page
Emberiza Diuca	92	*Fringilla Manimbè*	90
———— *Gayi*	92	———— *Magellanica*	97
———— *oryivorus*	106	———— *luteoventris*	88
———— *luteoventris*	88	Fringillidæ	87
———— *luctuosa*	94	Fringillinæ	90
———— *jacarina*	92	Fulica galeata	133
———— *gubernatrix*	88	*Furnarius Chilensis*	67
———— *cristata*	88	———— *Lessonii*	67
———— *cristatella*	88	———— *cunicularius*, Anat. Descript. of	148
Emberizina	88	———— *fuliginosus*	67
Emberizoides poliocephalus	98	———— *rufus*	64
Eremobius	69	———— *cunicularius*	65
———— *phœnicurus*	69	———— *dumetorium*	66
Etourneau des terres	110	———— *ruber*	80
Euphone jacarina	92	*Gafarron*	97
Falconidæ	9	Gallinula crassirostris	133
Falconina	28	———— *galeata*	133
Falco sparverius	29	Geospiza magnirostris	100
Falco femoralis	28	———— *strenua*	100
———— *degener*	13	———— *parvula*	102
———— *leucurus*	15	———— *nebulosa*	101
———— *Novæ Zealandiæ*	15	———— *fortis*	101
———— *Australis*	15	———— *crassirostris*	103
———— *Brasiliensis*	9	———— *dentirostris*	102
———— *histrionicus*	50	———— *fuliginosa*	101
Falcunculus Guianensis	58	———— *dubia*	103
Figulus albogularis	64	Grallatores	125
Fluvicolinæ	51	*Grande Mouette*	142
Fluvicola nengeta	54	*Grive rousse et noirâtre*	59
———— *icterophrys*	53	———— *blanche et noirâtre*	59
———— *Azaræ*	53	*Guirayetupa*	51
———— *Irupero*	50	Gyratones	114
———— *perspicillata*	53	Hæmatopus palliatus	128
Fournier	64	Halcyonidæ	41
Fregata Aquila	146	Halcyon erythrorhyncha	41
Fringilla Gayi	92	*Haliaëtus chimachima*	13
———— carbonaria	94	———— *erythronotus*	26
———— campestris	89	———— *chimango*	14
———— formosa	92	Hiaticula Azaræ	127
———— fruticeti	94	———— semipalmata	128
———— icterica	97	———— trifasciatus	127
———— *Hispaniolensis*	95	Himantopus nigricollis	130
———— matutina	91	Hirundinidæ	37
———— alaudina	94	*Hirundo purpurea*	38
———— splendens	92	———— concolor	39
———— Diuca	92	———— leucopygia	40

y 2

INDEX TO THE SPECIES.

	Page		Page
Hirundo frontalis	40	Mareca Bahamensis	135
—— cyanoleuca	41	Megalonyx medius	72
Huppe jaune	88	—— rufus	71
Hylactes Tarnii	70	—— albicollis	72
Ibis melanops	128	—— ruficeps	70
—— Ordi	129	—— rubecula	73
—— Falcinellus	129	—— rufogularis	73
Icterus fringillarius	107	Megalopterus stolidus	145
—— niger	107	Melanocorypha cinctura	87
—— anticus	107	Merops rufus	64
—— maxillaris	107	Micropterus brachypterus	136
—— sericeus	107	Milvago pezoporos	13
—— unicolor	107	—— montanus	19
—— sulcirostris	107	—— ochrocephalus	13
Irupero	53	—— chimango	14
Ispida torquata	42	—— megalopterus	21
Laniadæ	58	—— leucurus	15
Laniagra Guianensis	58	—— albogularis	18
Lanianæ	58	Mimus Patagonicus	60
Lanius doliatus	58	—— Orpheus	60
—— nengeta	54	—— parvulus	63
—— sulphuratus	43	—— Thenca	61
Larus fuliginosus	141	—— saturninus	60
—— hæmatorhynchus	142	—— melanotis	62
—— dominicanus	142	—— trifasciatus	62
—— cirrocephalus	142	Molothrus niger	107
—— maculipennis	142	Motacilla Patagonica	67
—— glaucodes	142	—— Gracula	67
Leistes erythrocephala	109	Mouette cendrée	142
—— anticus	107	Muscicapa psalura	51
Leptonyx macropus	71	—— risoria	51
—— Tarnii	70	—— mæsta	53
—— albicollis	72	—— nivea	53
—— paradoxus	73	—— parulus	49
—— rubecula	73	—— icterophrys	53
Lessonia erythronotus	84	—— pyrope	55
Lichenops erythropterus	52	—— polyglotta	54
—— perspicillatus	51	—— Tyrannus	43
Limosa Hudsonica	129	—— vittiger	54
Limnornis	80	Muscicapidæ	42
—— rectirostris	80	Muscipeta albiceps	47
—— curvirostris	81	Muscisaxicola brunnea	84
Lindo bleu dore et noir	97	—— nigra	84
Macas cornu	137	—— mentalis	83
Malacorhynchus Chilensis	73	—— macloviana	83
Manimbé	90	Muscivora Tyrannus	34

INDEX TO THE SPECIES.

	Page		Page
Myiobius parvirostris	48	Palmipedes	134
——— magnirostris	48	Parulus ruficeps	79
——— albiceps	47	Passerina discolor	107
——— auriceps	47	——— jacarina	92
Nothura minor	119	——— guttata	94
——— perdicaria	119	Passer Hispaniolensis	95
——— major	119	——— Jagoensis	95
Numenius Hudsonicus	129	Patagon	113
——— brevirostris	129	Patagonian maccaw	113
Nycticorax violaceus	128	——— warbler	67
——— Americanus	128	Pelecanoides Berardi	138
Œnanthe perspicillata	51	——— Garnotii	139
L'Ouglet	97	Pelecanus carunculatus	145
Opetiorhynchus rufus	64	——— aquilus	146
——— antarcticus, Anat. Descript.		Pelidna cinclus	131
of	149	——— minutilla	131
——— lanceolatus	68	——— Schinzii	131
——— nigro-fumosus	68	Le Pepoaza proprement dit	54
——— antarcticus	67	Pepouza variegata	55
——— Patagonicus	67	——— pyrope	55
——— vulgaris, Anat. Descript. of	149	——— maritima	57
——— Patagonicus, Anat. Descript.		——— gutturalis	56
of	150	——— nivea	58
——— vulgaris	66	Perdix Falklandica	117
——— rupestris	67	Perruche	112
Oreophilus totanirostris	125	Perspicilla leucoptera	51
Oriolus flavus	107	Petit Bout-de-Petum	114
——— cayannensis	106	Petrel échasse	141
——— ruber	109	Phalacrocorax imperialis	145
Ornismya Kingii	110	——— carunculatus	145
——— tristis	111	Phaleobœnus montanus	19
Orpheus Thenca	61	Philomachus Cayanus	127
——— melanotis	62	Phytotoma Bloxami	106
——— trifasciatus	62	——— rutila	106
——— Patagonicus	60	——— silens	106
——— modulator	60	——— rara, Anat. Descript. of	153
——— parvulus	63	——— rara	106
——— calandria	60	Picasuro	115
Ortyx Falklandica	117	Picus campestris	113
Otus Galapagoensis	32	——— Chilensis	114
——— palustris	33	——— Kingii	113
Oxyurus ornatus	81	——— melanocephalus	113
——— dorso-maculatus	82	Pigeon rougeatre	116
——— tupinieri	81	Pipillo personata	98
Pachyramphus albescens	50	Pitylus superciliaris	97
——— minimus	51	Platyurus niger	74

INDEX TO THE SPECIES.

	Page		Page
Podiceps kalipareus	136	Pyrrhulauda nigriceps	87
——— Rollandii	137	Pyrrhulinæ	88
——— Chilensis	137	Querquedula creccoïdes	135
Pœcilonitta Bahamensis	135	——— erythrorhyncha	135
Polyborinæ	9	Rallus Philippensis	133
Polyborus Galapagoensis	23	——— ypecaha	133
——— chimango	14	——— sanguinolentus	133
——— Brasiliensis	9	Recurvirostra himantopus	130
——— vulgaris	9	Regulus omnicolor	86
——— albogularis	18	——— Byronensis	86
——— chimachima	13	Rhea Americana	120
Porphyrio simplex	133	——— Darwinii	123
Prion vittatus	141	Rhinomya lanceolata	70
Procellaria oceanica	141	Rhynchaspis maculatus	135
——— Capensis	140	Rhynchæa semicollaris	131
——— Berardi	138	——— Hilairea	131
——— vittata	141	——— occidentalis	131
——— urinatrix	139	Rhynchops nigra	143
——— gigantea	139	Rhynchotus fasciatus	120
——— glacialoïdes	140	——— rufescens	120
——— puffinus	137	Sarcoramphus gryphus	3
Progne purpurea	38	——— Condor	3
——— modesta	39	Saurophagus sulphuratus	43
Psarocolius sericeus	107	Sauteur	92
——— anticus	107	Scansores	112
——— chrysopterus	106	Scolopax melanoleuca	130
——— flaviceps	107	——— vociferus	130
Psittacara Patachonica	113	——— Magellanicus	131
Psittacus Patagonus	113	——— Hudsonica	129
——— murinus	112	——— Paraguaiæ	131
Pteroptochos albicollis	72	——— Brasiliensis	131
——— albicollis, Anat. Descript. of	151	Scytalopus fuscus	74
——— Tarnii	70	——— Magellanicus	74
——— Tarnii, Anat. Descript of	150	Serpophaga	48
——— rubecula	73	——— parulus	49
——— megapodius	71	——— nigricans	50
——— paradoxus	73	——— albo-coronata	49
Ptiloleptus cristatus	114	——— albo-coronata, Anat. Descript. of	147
Puffinuria Berardi	138	Siuriri noirâtre et jaune	53
——— Garnotii	139	Sourciroux	58
Puffinus cinereus	137	Spermophila nigrogularis	88
Pyrocephalus	44	Spheniscus Humboldtii	137
——— parvirostris	44	Sphenura ruficeps	79
——— obscurus	45	Squatarola fusca	126
——— nanus	45	——— cincta	126
——— dubius	46	Sterna aranea	145
Pyrgita Jagoensis	95	——— stolida	145

INDEX TO THE SPECIES.

	Page		Page
Strepsilas interpres	132	Tanagra striata	97
Strigidæ	31	Tanagrinæ	97
Striginæ	34	Tantalus Mexicanus	129
Strix cunicularia	31	—— chalcopterus	120
—— flammea	34	Tetrao Falklandicus	117
—— punctatissima	34	Thalassidroma oceanica	141
—— brachyota	33	Thamnophilinæ	58
—— rufipes	34	Thamnophilus doliatus	58
Sturnella rubra	109	Theristicus melanops	128
—— militaris	110	Tinamus minor	119
Sturnus pyrrhocephalus	109	—— major	119
—— militaris	110	—— rufescens	120
Surninæ	31	Tinnunculus Sparverius	29
Sylvia Magellanica	74	Tinochorus rumicivorus	117
—— velata	87	—— rumicivorus, Anat. Descript. of	155
—— Bloxami	49	—— Eschscholtzii	117
—— macloviana	83	Tisserin des Galapagos	105
—— nigricans	50	Tityranæ	50
—— Patagonica	67	Tænioptera variegata	55
—— dorsalis	84	Totanus semicollaris	131
—— perspicillata	51	—— fuliginosus	130
—— rubrigastra	86	—— flavipes	129
Sylvicola aureola	86	—— macropterus	129
Synallaxis humicola	75	—— melanoleucos	130
—— major	76	—— solitarius	130
—— rufogularis	77	Trichas canicapilla	87
—— maluroides	77	—— velata	87
—— maluroides, Anat. Descript. of	152	Tringa Morinellus	132
—— flavogularis	78	—— minutilla	131
—— brunnea	78	—— Urvillii	126
—— ægithaloides	79	—— rufescens	130
—— ruficapilla	79	—— semipalmata	128
—— dorso-maculata	82	—— macroptera	129
—— tupinieri	81	—— Schinzii	131
Tachuris omnicolor	86	Trochilidæ	110
—— roi	86	Trochilus flavifrons	110
—— nigricans	50	—— forficatus	110
Le Petit Tachuris noirâtre	50	—— gigas	111
Tanagra ruficollis	91	—— gigas, Anat. Descript. of	154
—— Guianensis	58	Troglodytes Magellanicus	74
—— vittata	98	—— Platensis	75
—— canicapilla	87	—— paradoxus	73
—— Darwinii	97	Troupiale à tête jaune	107
—— Bonariensis	107	Turdidæ	59
—— superciliaris	97	Turdus Falklandicus	59
—— jacarina	92	Turdus Magellanicus	59

INDEX TO THE SPECIES.

	Page		Page
Turdus Thenca	61	Vanellus cinctus	126
——— rufiventer	59	Viralva aranea	145
——— *varius*	64	*Vultur gryphus*	3
——— *curæus*	107	——— *aura*	8
——— *Chochi*	59	——— *atratus*	7
——— *leucomelas*	59	——— *jota*	7
——— *albiventer*	59	Vulturidæ	3
——— *Orpheus*	60	White rumped snipe	130
Tyranninæ	42	Xanthornus chrysopterus	106
Tyrannula magnirostris	48	——— flavus	107
——— *auriceps*	47	Xema cirrocephalum	142
——— *parvirostris*	48	Xolmis nengeta	54
Tyrannus Irupero	53	——— coronata	54
——— *pepoaza*	54	——— variegata	55
——— *polyglottus*	54	——— Pyrope	55
——— *nengeta*	54	*Yahana proprement dit*	133
——— *magnanimus*	43	*Ypacaha*	133
——— *sulphuratus*	43	*Yellow crested grosbeak*	88
——— *gutturalis*	56	*Yellow shanks snipe*	129
——— *coronatus*	54	*Yetapa psalura*	51
——— *Savana*	43	Zaporina notata	132
Ulula rufipes	34	——— spilonota	132
Ululinæ	32	Zenaida aurita	115
Uppucerthia dumetoria	66	——— Galapagoensis	115
——— dumetoria, Anat. Descript. of	148	——— Boliviana	116
——— *dumetorum*	66	Zonotrichia strigiceps	92
Uppucerthia vulgaris	66	——— canicapilla	91
——— *rupestris*	67	——— matutina	91
Uppucerthia nigro-fumosa	68		

Milvago albogularis.

Craxirex Galapagoensis.

Otus Galapagoensis.

Strix punctatissima

Progne modestus.

Pyrocephalus parvirostris.

Pyrocephalus nanus.

Tyrannula magnirostris.

Lichenops erythropterus.

Fluvicola Azaræ

Birds Pl. 11

Tænioptera variegata.

Agriornis micropterus.

Agnornus leucurus

Pachyramphus albescens

Pachyramphus minimus.

Mimus trifasciatus.

Mimus melanotis.

Mimus parvulus.

Upercerthia dumetaria.

Opetiorhynchus lanceolatus

Eremobius phœnicurus.

Synalaxis rufogularis.

Synalaxis flavogularis.

Birds Pl. 25.

Limnornis curvirostris

Limnornis rectirostris.

Dendrodramus leucosternus

Sylvicola aureola

Ammodramus longicaudatus.

Ammodramus xanthornus.

Passer Jagoensis

Chlorospiza melanodera

Chlorospiza Xanthogramma

Tanagra Darwini

Pipilo personata.

Geospiza magnirostris

Geospiza strenua.

Geospiza fortis

Geospiza parvula.

Camarhynchus psittaculus.

Camarhynchus crassirostris.

Cactornis scandens.

Birds Pl. 43.

Cactornis assimilis.

Certhidea olivacea.

Xanthornus flaviceps

Zenaida Galapagoensis.

Birds Pl. 47.

Rhea Darwinii

Birds Pl. 48

Zapornia spilonota.

Printed in May 2022
by Rotomail Italia S.p.A., Vignate (MI) - Italy